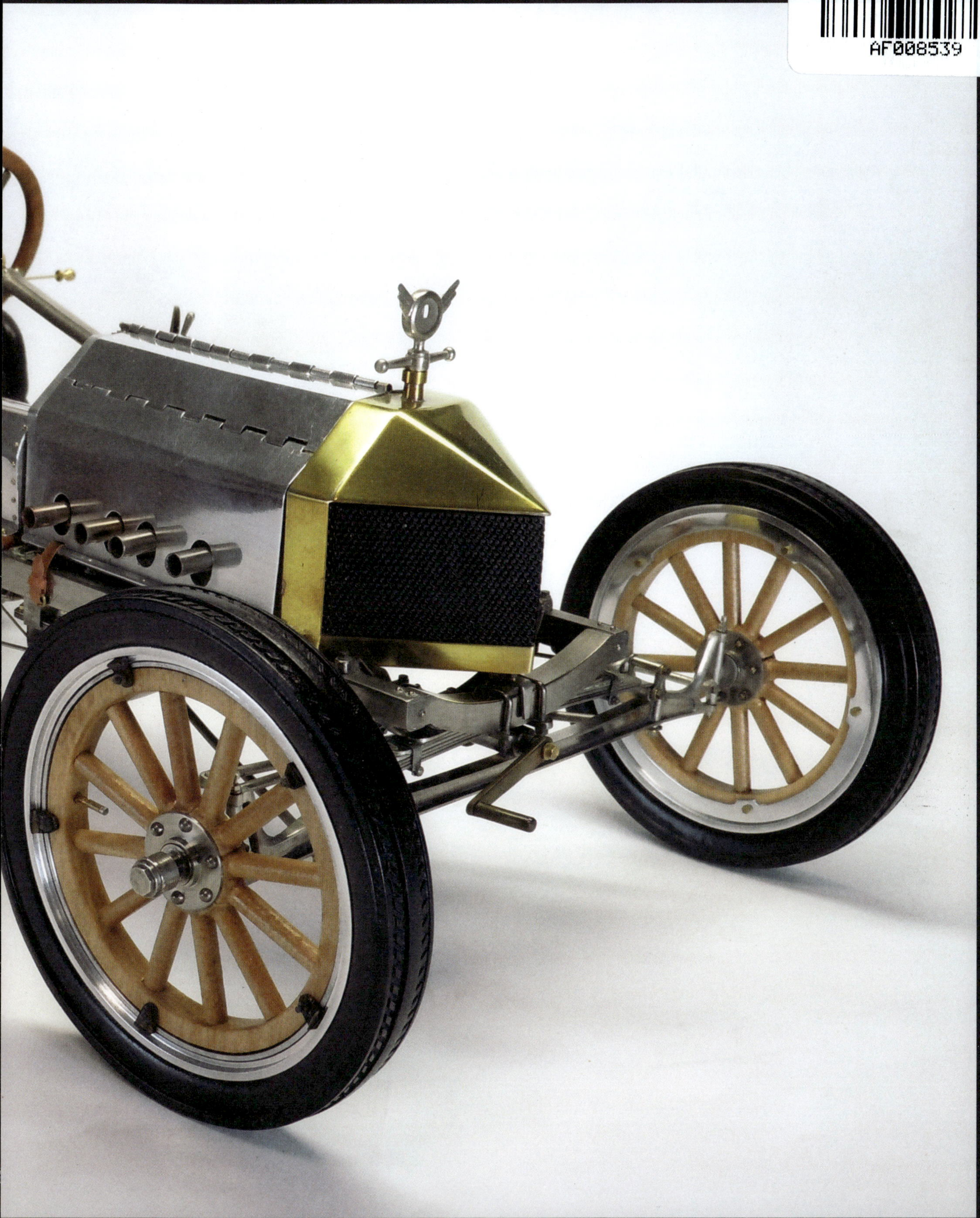

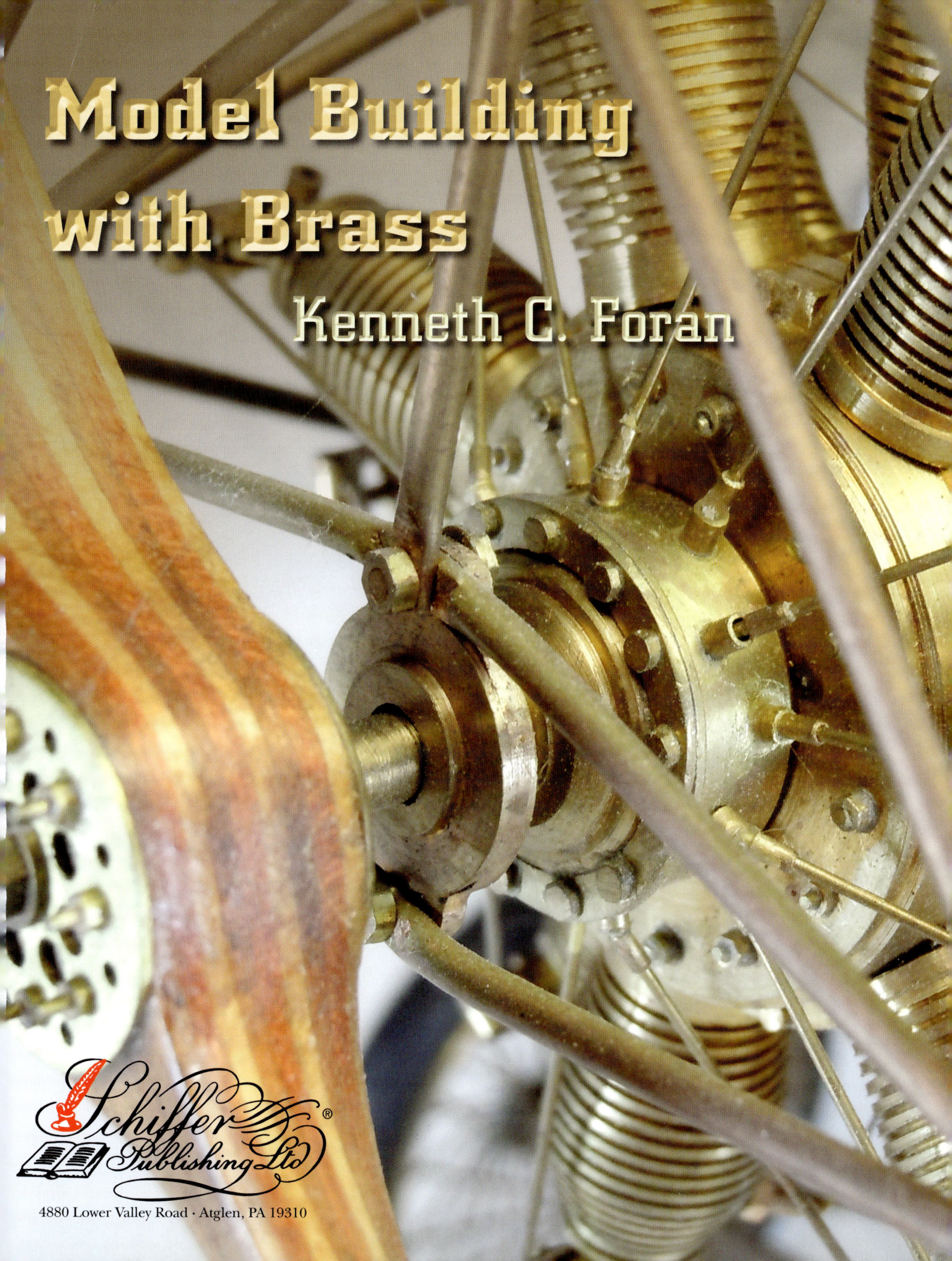

Model Building with Brass

Kenneth C. Foran

Other Schiffer Books on Related Subjects:
Fundamentals of Model Boat Building. John Into & Nancy Price. ISBN: 9780764331053. $34.99
The Master Scratch Builders: Tips & Techniques from the Master Aircraft Modelers. John Alcorn. ISBN: 0764307959. $39.95
Model Boat Building: The Skipjack. Steve Rogers & Patricia Staby-Rogers. ISBN: 0887409377. $14.95
Ship Model Building. Gene Johnson. ISBN: 9780870333699. $12.95

Copyright © 2012 by Kenneth C. Foran

Library of Congress Control Number: 2011944426

All rights reserved. No part of this work may be reproduced or used in any form or by any means—graphic, electronic, or mechanical, including photocopying or information storage and retrieval systems—without written permission from the publisher.
The scanning, uploading and distribution of this book or any part thereof via the Internet or via any other means without the permission of the publisher is illegal and punishable by law. Please purchase only authorized editions and do not participate in or encourage the electronic piracy of copyrighted materials.
"Schiffer," "Schiffer Publishing Ltd. & Design," and the "Design of pen and inkwell" are registered trademarks of Schiffer Publishing Ltd.

Type set in Acanthus/Zurich BT

ISBN: 978-0-7643-4004-8
Printed in China

Schiffer Books are available at special discounts for bulk purchases for sales promotions or premiums. Special editions, including personalized covers, corporate imprints, and excerpts can be created in large quantities for special needs. For more information contact the publisher:

Published by Schiffer Publishing Ltd.
4880 Lower Valley Road
Atglen, PA 19310
Phone: (610) 593-1777; Fax: (610) 593-2002
E-mail: Info@schifferbooks.com

For the largest selection of fine reference books on this and related subjects, please visit our website at
www.schifferbooks.com
We are always looking for people to write books on new and related subjects. If you have an idea for a book, please contact us at
proposals@schifferbooks.com

This book may be purchased from the publisher.
Include $5.00 for shipping.
Please try your bookstore first.
You may write for a free catalog.

In Europe, Schiffer books are distributed by
Bushwood Books
6 Marksbury Ave.
Kew Gardens
Surrey TW9 4JF England
Phone: 44 (0) 20 8392 8585; Fax: 44 (0) 20 8392 9876
E-mail: info@bushwoodbooks.co.uk
Website: www.bushwoodbooks.co.uk

Dedication

This book is dedicated to the little kid in each of us who enjoys building models, no matter how old we are.

Ken Foran

ACKNOWLEDGMENTS

First, I would like to thank my wife, Gretchen, and my daughter, Heather, for their patience and editorial help with the writing of this book. Also, I'd like to thank those online model builders around the world that encouraged me and suggested I write this book describing the techniques of building with brass.

Contents

ABOUT THE AUTHOR — 6
INTRODUCTION — 12
CHAPTER 1: WHY USE BRASS & BRASS STOCK — 13
CHAPTER 2: VARIOUS TOOLS — 15
CHAPTER 3: WORKING WITH BRASS — 25
   Round Tubing — 25
   Profiles, Shapes, and Rods — 31
   Sheet Stock — 33
   Model Building — 35
   Soldering — 35
   Soldering Irons — 35
   Solder — 35
   Fluxes — 36
   Soldering Operations — 36
CHAPTER 4: PART FABRICATION – USING SHEET STOCK — 43
   Part Fabrication: Incorporating Round Tube — 55
   Part Fabrication: Various Stocks — 67
   Part Fabrication: Using Bar Stock — 71
   Part Fabrication: Small Parts — 76
   Part Fabrication: Using Laminations — 83
   Using Binding Wire — 92
CHAPTER 5: INCORPORATING COPPER WITH BRASS — 93
   Using Other Modeling Materials — 101
   Making a Wire Wheel — 104
   Working with Wire — 106
CHAPTER 6: MICRO HARDWARE — 108
   Taps — 108
   Dies — 110
   Nuts, Bolts, and Machine Screws — 111
CHAPTER 7: CLEANING PARTS — 112
CHAPTER 8: TABLETOP MACHINING CAPABILITIES — 114
   Drill Press — 114
   Tabletop Lathe — 115
   Tabletop Milling Machine — 119
CHAPTER 9: WORKING WITH CHEMICALS — 120
   Cleaning — 120
   Electroplating — 120
   Photo Etching — 125
   Patinas — 126
CHAPTER 10: WHAT TO DO WHEN YOU MESS UP — 127
CHAPTER 11: BRINGING IT ALL TOGETHER — 128
CHAPTER 12: SOURCES AND SUPPLIERS — 160
   Places to Visit — 160

ABOUT THE AUTHOR

Ken Foran was born in Marathon, Ontario, Canada, immigrated to the United States in 1965, and enlisted in the Marine Corps for the GI Bill to attend college. Upon completing his three-year enlistment from 1965-1968, including a 13-month tour in Vietnam, he attended the Cleveland Institute of Art and graduated with a B.F.A. in Industrial Design. Ken has spent his entire working career in Product Design and Development. He and his development teams have been awarded numerous patents and design awards, both domestically and internationally. Under his responsibility, Ken had a product development model shop that utilized state-of-the-art prototype technology.

Throughout his youth, he built various plastic and wood models of airplanes, cars, and ships. Early on, Ken would go beyond the normal kit, doing what now has been coined as "kit bashing." In college, his building skills evolved to developing the prototype models of the products he designed. Ken was selected to be one of the four-man team of the 1970 Clean Air Car Competition representing the Cleveland Institute of Art. Ken was the only underclassman on the team that built a fully-functional prototype car in six months that was tested at the GM Proving Grounds in Milford, Michigan. The car was awarded the Best Design Award. Ken's background and education in the Marines are what really stood out as an asset in building the car. Ken was trained as a Helicopter Structural Mechanic and transferred his skills from helicopters to the car build.

While his working career was very demanding of time and he traveled extensively around the world, he managed to work with both of his children, Eric and Heather, on various school projects that required building. His son figured out early on that

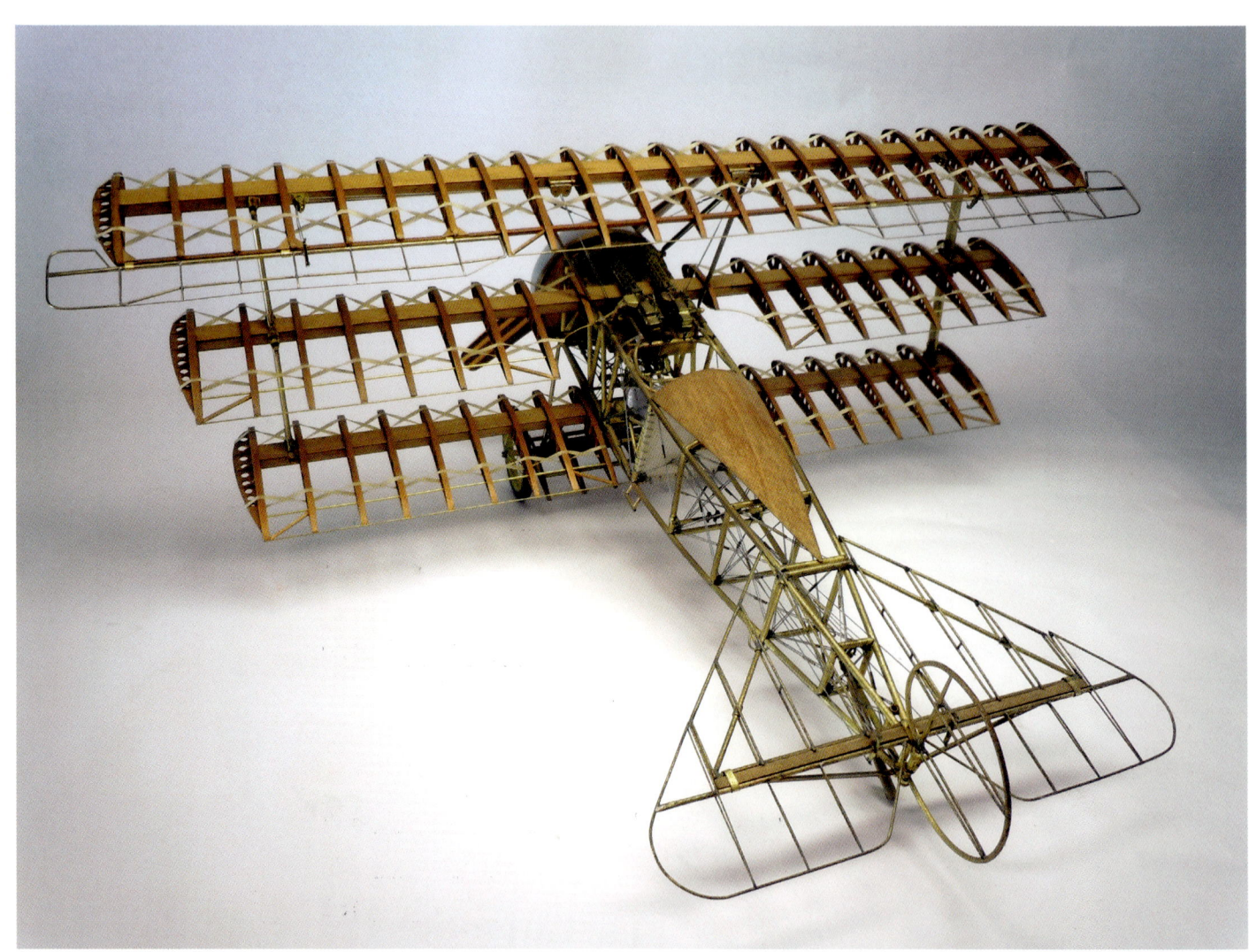

he preferred flying radio controlled planes, instead of building them. Ken was a better builder than a flyer, so they made a great team.

Over the years, Ken accumulated an extensive collection of hand tools and tabletop machines found at local tool sales. Ken admits to not being a machinist and is self-taught on the equipment. A lesson taught to him by his father was that sometimes the biggest mistake you can make is to buy cheap tools. He also taught himself how to use computers and software programs over the years.

Ken has had a lifelong interest in flying and airplanes, especially the WWI era planes. He attends the Dawn Patrol fly-ins at the Wright Patterson Air Force Base held biannually.

Ken, after years of assembling model kits and being disappointed with the quality and level of details, decided to take matters into his own hands and started scratch building. His first attempt was a Fokker Triplane. Ken does all of his own research and scours the Internet looking for resources, reference material, and images of the planes to build. He is careful about using the drawings of others because he has found errors or outright mistakes in some drawings. Whenever possible, he obtains original reference drawings or pictures from the period.

Ken scratch-built the Fokker Triplane, which was his first attempt and learning experience. His second project was a Fokker D VII. He selected this plane because of the engine. Ken taught himself how to solder, fabricate, and machine brass. Brass is by far the fastest and easiest medium to work and build with. "Many times I can build component parts while they are still hot from soldering, not having to wait for glue to set." With brass, just about any shape, size, profile configuration or sheet thickness is available and can be soldered together. The Fokker D VII Mercedes engine did in fact challenge his building skills; he hand-fabricated the upper and lower crankcases using a layering technique. The final shaping was completed with files. He had to work very slowly and carefully because there was no chance of using filler if a mistake was made.

As with all of Ken's models, they are built as close as possible to the real subject. All control surfaces work as they should from the joystick and rudder bar. In some cases, they even have working suspensions. Gretchen, his wife, helps by making the cowls when needed. Gretchen is also a graduate of the Cleveland Institute of Art in Silversmithing and taught there for ten years. Ken says it is amazing to watch her take a small circle of flat aluminum and, with nothing more than hammers and stakes, hammer the cowls not only round but also to a press-fit tolerance.

Ken posted images of his two Fokker planes on a popular WWI website and immediately was noticed. Ken started his next project, a Fokker Eindecker E-IV, to round out the Fokker series with an obscure model with a double bank rotary engine. Gary Kohs, of Fine Art Models, emailed and invited Ken to visit him in Detroit to see Ken's work in person.

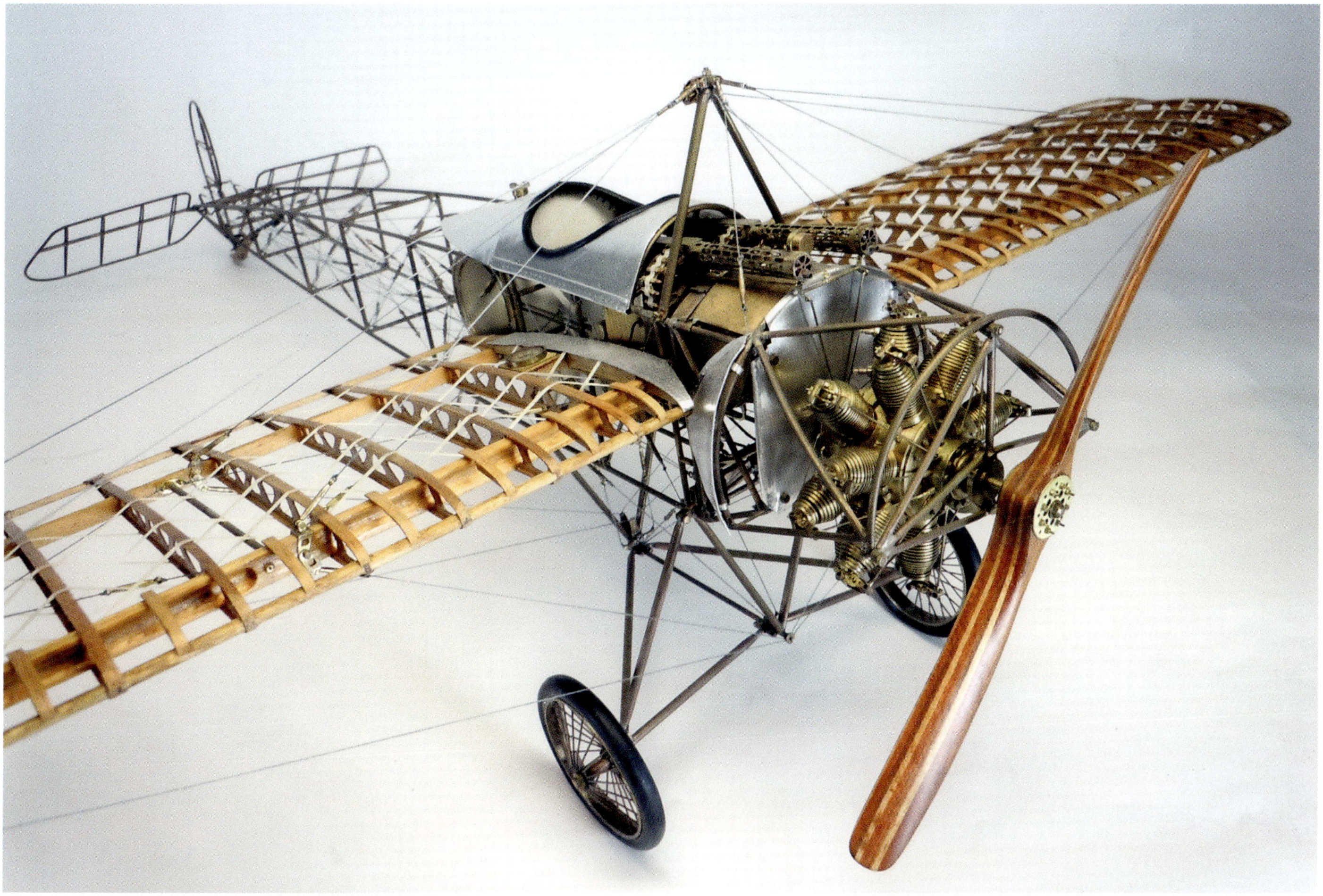

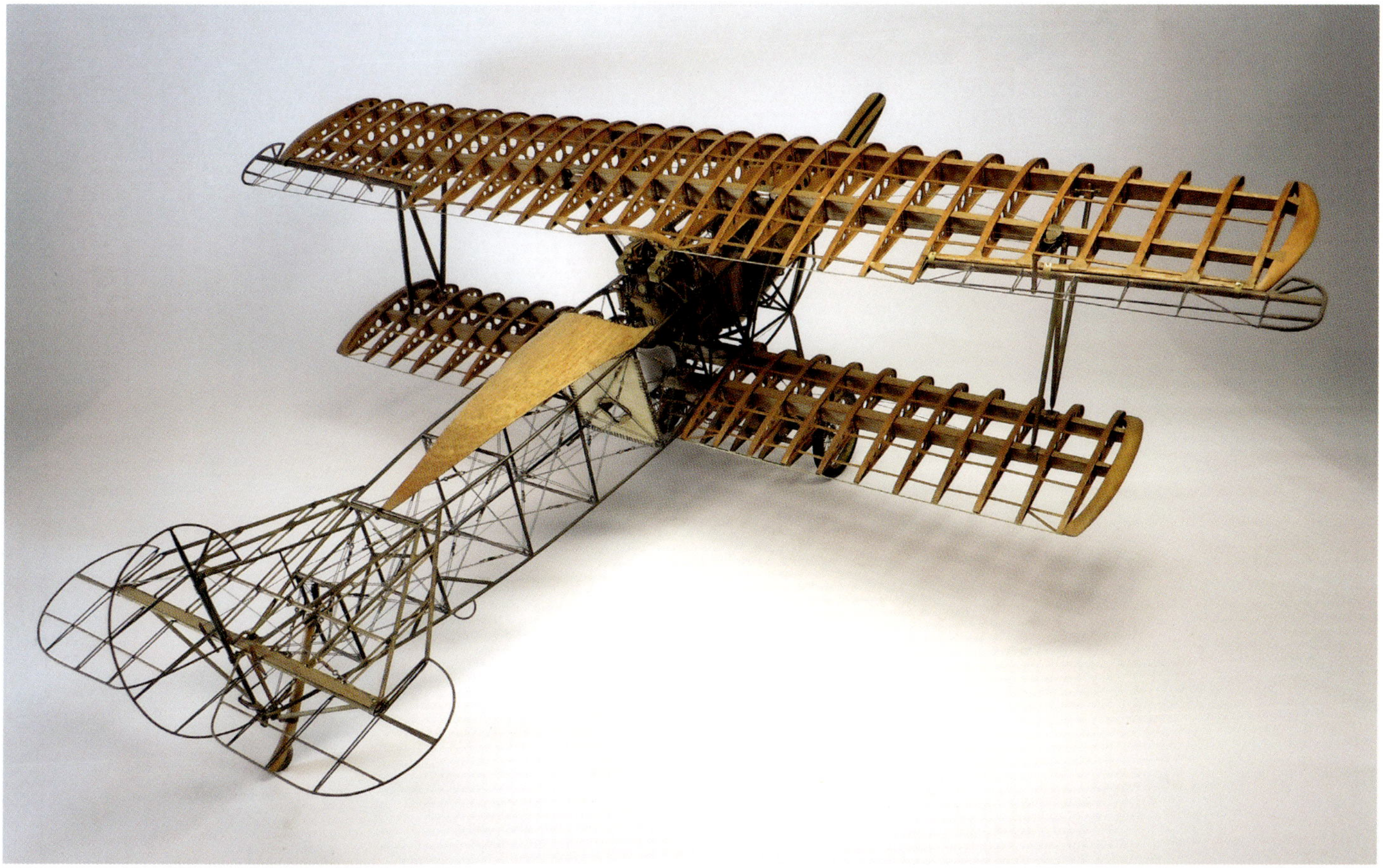

Within five minutes of looking at the models, Gary invited Ken to attend the upcoming International Toy Fair in Nuremberg, Germany. He commissioned Ken to build a Sopwith Camel for Fine Art Models. Upon the completion of the Sopwith Camel in 1/15th scale, and exceeding Gary's expectation with the build, Gary then offered Ken a challenge.

Gary told Ken he always wanted to build a certain model, but in the previous thirteen years he had not been able to find a builder to do it. Much to Ken's surprise, it was a Bell H-13D Sioux M.A.S.H. helicopter. With Ken's background as a helicopter mechanic in the Marines, this idea got his attention. Ken accepted the challenge and located an H-13D being restored in Canada. He visited with them, shooting 10 rolls of film, and made numerous sketches with dimensions. Ken then located and purchased original parts catalogs and erection manuals with more details and dimensions. Ken tries to be as accurate as possible when he builds a model, and the better the information, the more accurate the model.

The challenge building the H-13D was that most of the components had to be fabricated out of brass. The model was not only a prototype but was also to be used to make the master patterns and molds for production of the H-13D. This required the model to be able to be disassembled, then reassembled. To add features, Ken also geared the main rotor to turn the tail rotor and engine cooling fan. There was little margin for error because every component was visible.

Fine Art Models exhibited the H-13D model at the London Model Engineering Show in London, England, and at the International Toy Fair in Nuremberg, Germany. Several of Ken's WWI planes were exhibited at previous International Toy Fairs. Ken's models are also included in the Internet Craftsmanship Museum hosted by Sherline, Inc.

Overall right-side view.

Introduction

I built my first small plastic Viking ship model at the age of ten in 1957. As the years went by, I conquered more and more complex kit models. I was encouraged by my grandfather, who hand carved clipper ship models. Over time, my skills improved and a greater desire for more intricate details evolved.

Standard model kits were that—standard. What evolved were cars, planes, and boats, all modified just as the hot rods were in that period of time. But, cast white metal and plastic have structural limitations. After watching an expensive $1/16^{th}$ scale WWI airplane model splay its landing gear and collapse under its own weight, I tried a new substitute material – BRASS.

Brass stock was readily available at a local hardware store in sizes, shapes, and profiles that could be used in model building. So, I rebuilt the collapsed landing gear with brass. I found the brass to be easy to cut, shape, and solder. The structure was much stronger than the kit supplied parts and the resulting finished surfaces were also smoother and more refined.

In writing this book, I will attempt to share what I've learned and experienced in building models with brass.

May your brass built models be strong ... let the building begin!
Ken

Chapter 1
Why Use Brass

Brass is essentially an alloy of copper and zinc in various proportions. Brass stock is commercially available in many various forms, shapes, and profiles. For the scope of this book, I will focus on sheet stock, bar, tubing, rod, and wire as they relate to building models.

In modeling, whether airplanes, cars, ships or structures, there is a time when one material is better suited for an application than another. The great benefit of brass is the commercial availability of many various profiles and shapes, but the real benefit is that it can be soldered quickly and easily while also providing a great strength to weight ratio. Brass can be easily worked with tools and files into just about any configuration imaginable. When finished and cleaned, it is easily painted. Also, brass is very forgiving if mistakes are made. After learning the techniques of working with brass and soldering, the benefit of speed becomes apparent when not having to wait for glues or adhesives to set.

Brass Stock

Many various forms, shapes, and dimensions of brass are commercially available. Following is a brief overview of the different types of brass stock to work with.

Sheet Stock: Sheet stock comes in various sizes, thicknesses, and hardness. For example, foils are extremely thin and have no structural integrity themselves but can be applied over an existing surface. Shim stocks used in the industrial manufacturing process to adjust dimensions come in many various thicknesses or gauges and are generally very hard. I will not bore you with the difference, but I recommend a good understanding of decimals, since most commercial sales are based on the decimal system for thickness. Many of the modeling experiences in fabrication that I have encountered can be addressed with .005", .010" or .015" sheet thickness; for example, on gas tanks in airplane models. However, there may be occasions for the need to use thicker stock. Soft brass must be used if you plan to form a shape over a mandrel or buck to obtain a form other than flat. For the purpose of this book, we will stick with the simple fabrication techniques and leave hammer forming to the metalsmith. But if you feel so inclined, by all means try it. A great resource book to consider is the ***The Complete Metalsmith*** by Tim McCreight, which explains, in simple terms, many of the processes involved.

Tubing: Tubing is probably the greatest profile that satisfies most common applications. Again, tubing comes in as many sizes as one can imagine. One great benefit many suppliers provide is the telescoping sizes. This means one size tube slips into another tube. In modeling and soldering, this telescoping feature is very important to duplicate some objects or joints. Always think of what part you are trying to replicate and if you need telescoping sizes. Tubing profiles come in round, square, hexagonal, rectangular, and a few elliptical shapes. In the common sizes, various wall thicknesses are available. This is where your understanding of decimals will come in handy. Greater wall thickness provides greater strength. In some cases, those who have lathes will need thicker wall stocks for turning purposes to replicate a part.

Bar Stock: Bar stock, like tubing, comes in various shapes, including round, square, rectangular, and hexagonal. Bar stock is what it implies. It is solid and suitable for cutting and shaping solid parts. Round bar stock under .025" in diameter is generally referred to as rod. Again, it is sold based on decimal sizes and various lengths, usually 12" or 36" long.

Wire: Wire generally is sold in two forms—on a roll or in lengths. Roll wire is generally a softer alloy than that sold in lengths, which is a half hard alloy. Roll wire is sold by using the gauge size; the more enlightened manufacturers also provide a decimal equivalent dimension. Both types have particular value in satisfying certain requirements. Rod lengths can slip inside tubing and be soldered in place to provide attachment points and simulate wiring or cabling. Roll wire can be wound around a rod or mandrel to simulate a spring or woven (which will be discussed in Chapter 5) to make a wicker seat.

Brass comes in various ranges of hardness. Most stock sold, unless otherwise noted, is generally what is referred to as "half hard." Most small tubing, rod, and sheet stock would be in this category. Generally, "soft" brass has to specifically be requested when ordering material. Soft brass is used to form three dimensional shapes that require being hammered over a form or stake. Soft tubing has an application for fuel lines in radio controlled vehicles and is malleable enough to form without kinking. A modeling application would be forming radiator hose or exhaust systems.

Half hard brass can be slightly softened through the annealing process by taking the piece to be annealed and heating it with a butane torch until the metal is a dull red color, then quench in water. This is opposite to ferrous metals (steel and iron), which would get harder. This process is true for most non-ferrous alloys. Annealing brass will bring the copper color to the surface. Polishing or sanding will take it back to the original brass color. Hammer formed parts that have work hardened can also be annealed for further forming. Complex forms can be annealed many times during the forming process.

Chapter 2
Various Tools

Having the right tool for the job is the key to the successful completion of a project. Sometimes the most expensive mistake you can make is to buy cheap tools. Cheap tools usually do not have the close tolerances needed when working at small scales. Over time, one will gather an assortment of hand tools. Some will be essential and used often while others will be nice to have for that special occasional need.

The basic operations in working brass are cutting, shaping or forming, soldering, cleaning, and then finishing.

Cutting: Brass is a non-ferrous alloy, which means, compared to steel, it is soft and easier to work. Cutting thin stock can be accomplished with an X-ACTO® knife, scissors, jeweler's saw or shears.

X-ACTO® Knife: The #11 general purposes blade will be one of the most useful tools in your toolbox. Purchase blades in bulk packs of 100 for the most cost-effective savings. Sometimes, this method can be the most efficient way to cut small tubing and rod. Practice over time will help set your limits, but I generally cut tubing and rod up .125" in diameter using the knife. Here is how to do it. First, you will need to get a small, flat piece of hardwood (oak) molding, door casing or floor molding that has the recessed back around 6 to 8 inches long. Take the tubing or rod and lay it perpendicular to the grain of the wood. Place the blade edge perpendicular to the tube and parallel to the wood surface and roll with a little pressure back and forth around the perimeter of the tube. A steel bench block can be used instead of the wood if you have one. A score line will appear; now, the trick is to increase pressure, stay in the cut groove, and continue to roll back and forth until the tube is cut. This takes a little practice to not wander out of the groove. Rod can be cut the same way, but not all the way through. Cut deep into the rod, and then it should break under slow, even bending force. This technique is replicating the same action as a pipe cutter does, only in miniature. After cutting, you will need to file clean the outside burr raised during the cutting, and the inside burr can be carefully removed with the X-ACTO® blade point around the inside of the tube perimeter. Very precise cutting and small dimensions can be accomplished with practice.

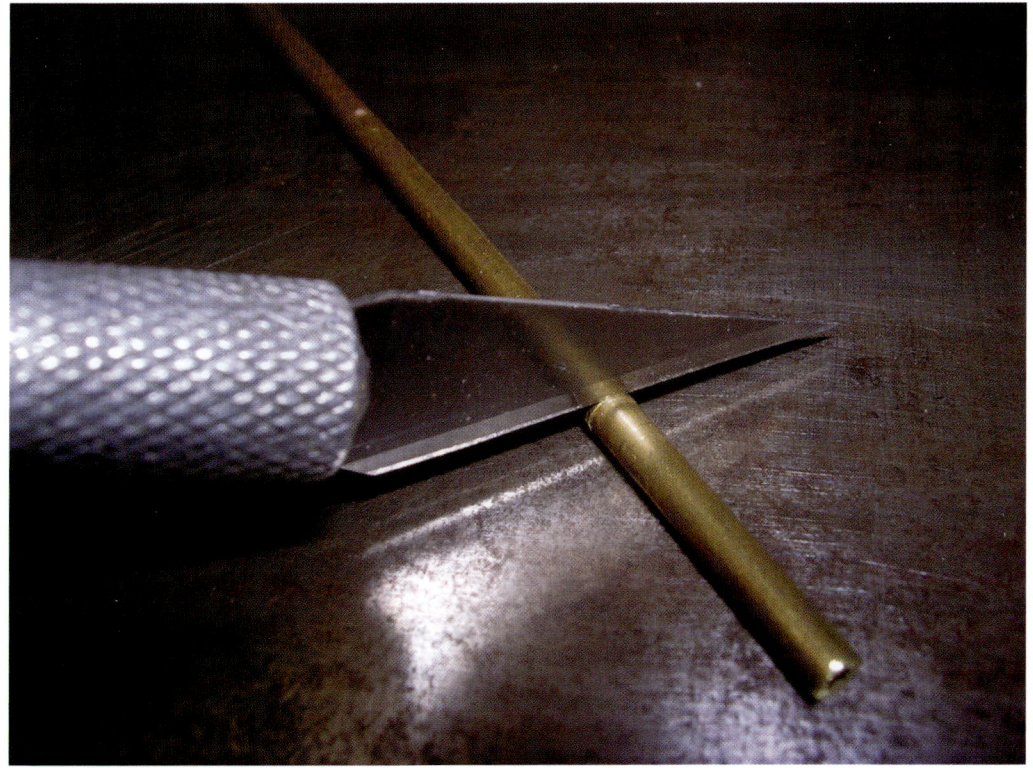

Jeweler's Saw: Invest in a good jeweler's saw and get an assortment of blade sizes. A jeweler's saw again is one of those essential tools you will need to get used to with practice. The fine metal blades will snap and break easily if not used correctly and one needs to develop the "feel" of cutting. Using beeswax as a lubricant on the blade is a must. Beeswax both lubricates while cutting and also reduces buildup of brass cuttings in the teeth of the blade, thus optimizing the cutting process while extending the life of the blade. This is a case when more is better; beeswax is much cheaper than blades. Blades come in sizes ranging from "0000" size on up. The most commonly-used sizes are 0, 1, and 2, and these can handle cutting .125" thick bar stock. While cutting, use a gentle vertical stroke and let the blade do the cutting. Forcing it will make the blade snap. When cutting intricate shapes out of brass sheet, I recommend using a jeweler's bench pin and spray glue the brass sheet to a sacrifice piece of wood stock. Large tubing is best cut with this saw if you do not have a lathe. Critically important to cutting any stock is to make sure the piece being cut is held firmly. When the jeweler's saw is not in use, release the pressure on the blade and store blades in a clean, dry location. Rusting of the teeth can cause the blades to snap when in use.

Bench Pin: The best tool to complement a jeweler's saw is what is referred to as a bench pin.

Use a jeweller's saw to cut out the member side shapes. Use beeswax (or an old used candle) to lubricate the saw blade and expect to break blades until you get the feel. Sawing on a bench pin is best. Note: wife has me trained to saw over a wastebasket.

Bench pins can be purchased or you can make your own, which is what I did out of a piece of oak .75" x 3.5" x 8" long. I cut a 2" deep "V" notch .75" in from each edge for a flat working surface. I then band saw cut a taper from the front edge back to the point of the notch. Then, an elongated hole was cut to allow room for a C-clamp to be inserted to clamp the pin to the work bench edge. If you have enough space, the bench pin can be permanently screwed to the bench.

Taper to front for clearance.

Scissors: Scissors can cut foils and sheet stock up to .020" very effectively. I recommend purchasing a good high-quality product that will be dedicated to this purpose. Borrowing the household pair is not a good idea, especially if there is a seamstress in the house. Scissors can be used for rough, as well as, fine cutting of shapes. Again, practice will develop skill of use and personal limits.

Laid out side rails on .016" brass sheet and cut out with sharp scissors. Use a scribe to get thin crisp lines to follow. Cut two identical.

Shears: Shears come in two varieties, hand held and bench, the use of which is dictated by the thickness of sheet or bar stock being cut. Hand shears, also called aviation snips, can cut simple compound shapes, depending upon the size of the shears and the size of the shape. Bench shears are generally intended for straight cuts. The most useful shear I have found is an antique typesetter's shear. This style is compact and has stops to set sizes, which are handy when cutting several items of the same size. If you are fortunate to find one of these in your travels, buy it because it will be one of your best long term investments.

Files: One can never have enough files. Files are used for shaping stock, dressing ends of tubes, and cleaning solder joints. Also have a good file card to clean your files. In working with brass, you can use small bastard files on down to fine tooth files in the 4" - 6" size range. A variety of coarse needle files and fine needle files that can be purchased in sets are a necessity. In some cases, you may even need Riffle files in medium and coarse.

To dress and clean off excess solder from joints, use extra course and course needle files. Solder builds up and fills the teeth, so cleaning files is a constant challenge. To help reduce the solder build up, apply talc powder to the file teeth. The talc powder acts as a release agent, making cleaning easier.

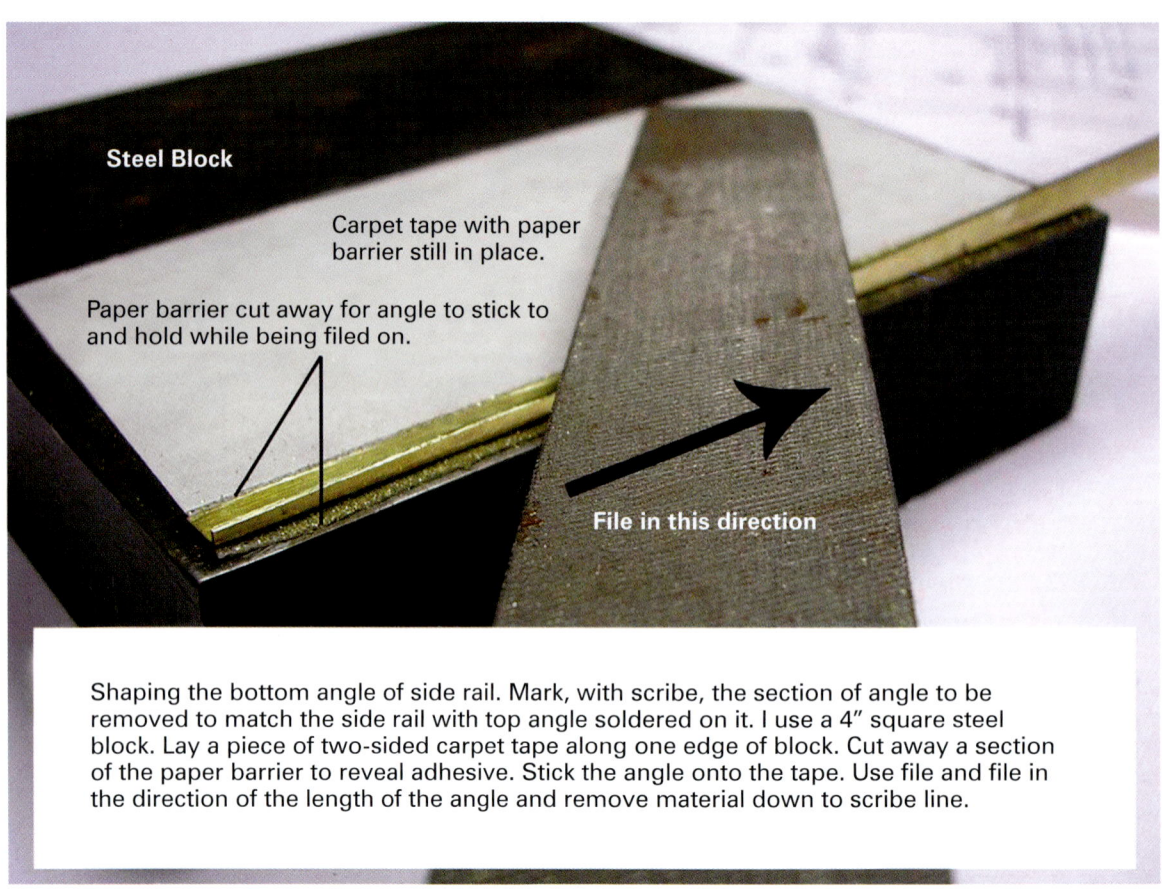

Shaping the bottom angle of side rail. Mark, with scribe, the section of angle to be removed to match the side rail with top angle soldered on it. I use a 4" square steel block. Lay a piece of two-sided carpet tape along one edge of block. Cut away a section of the paper barrier to reveal adhesive. Stick the angle onto the tape. Use file and file in the direction of the length of the angle and remove material down to scribe line.

Brushes: Small hand brushes are used to clean, scrub joints, and polish. Hand brushes that look like tooth brushes with wooden handles and have brass or steel bristles are the kind needed. The steel ones can also be used to clean small files. Small brushes of brass and steel for use with handheld rotary tools are also good to clean and polish finished solder joints. Make sure to wear safety glasses when working with any kind of rotary tool.

Soldering Irons: Soldering irons are the secret to successful soldering. Again, practice over time will enhance your skill level. The scale and thicknesses of the stock will dictate the proper wattage of the iron. Soldering irons are rated in watts from 25 and up. They can be single wattage units or adjustable in wattage settings. Based upon budget, I recommend a low 40 watts for small thin stock, a medium 60 watts for larger tubes or stock, and a high 120 watts for thick stock or multiple joints. In layering of heavy .125" stock, like in building engine blocks, a butane torch will be needed. Soldering iron tips come in various styles. I recommend round point and a chisel point for each size of iron. More on how and why will be explained later in the soldering section in Chapter 3.

Care and safe procedures must always be taken when soldering. *Soldering irons are hot!!!* Fingers were not made to be soldered.

Solder in a well-ventilated room; fumes can be harmful. **Keep children and pets away during soldering operations. Wear safety glasses!** Flux splatters when hot, so wear safety glasses. You will hear this again.

Picks: Picks are small pointed tools of stainless steel that are used during soldering operations to perform various tasks like holding parts in place and such. They are stainless steel because you cannot solder them to your model parts, which is a handy feature. However, some silver solders claim to be able to solder stainless steel, so check the solder instructions. These picks can be purchased in a set and are handy for other operations as well.

Tweezers: Tweezers are always handy to have—both the locking and non-locking kinds. Think of tweezers as extensions of your fingers that will not burn. The locking types are especially handy for holding small parts together during soldering. They are also great for picking up tiny bolts, washers, and nuts.

Pin Vise: Pin vises are the best tools to hold small drill bits for both machine and hand drilling operations. They come in various styles and capacities; some go down to zero while others do not close all the way, so check their capacities when purchasing. Pin vises are valuable when tapping the small sizes of threads with taps smaller than 00-90. Pin vises can also hold small parts while working on them.

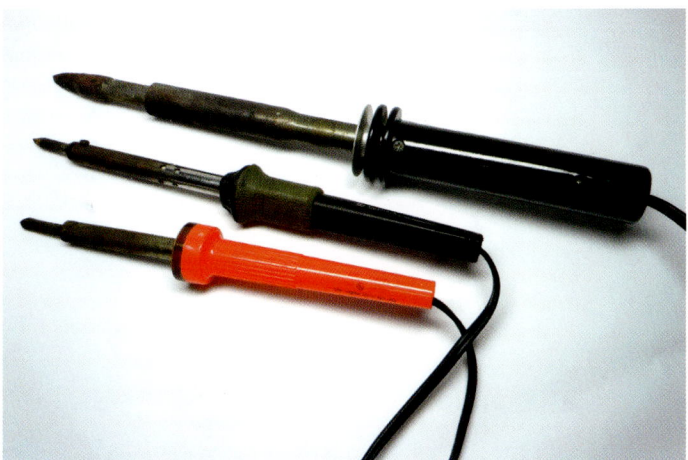

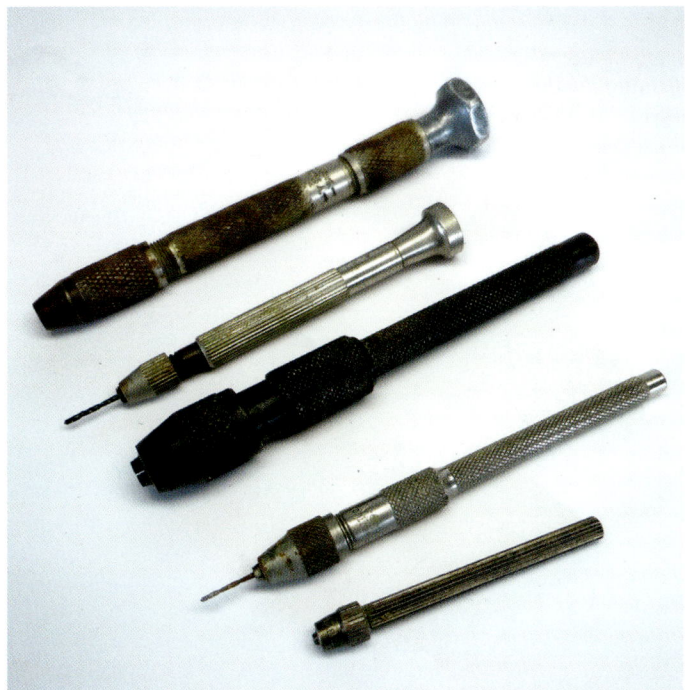

Reamers: A reamer is a tapered, hardened steel square shaft with a ground square end that fits into the working handle. Purchase reamers in a set of assorted sizes. When trying to get a tight press fit of a tube or rod in a hole, drill undersize and use a reamer to sneak up on the press fit. Insert the reamer into the hole until engaging the brass, then carefully twist slowly to remove material a little at a time, then test fit into the hole. Work slowly so as not to make the hole too large. Be careful with the very small sizes so as not to break them—being hardened they break very easily. Sets usually come in a plastic case; make it a practice to use them and keep them clean and replace back in the case immediately after use. Being ground steel, they are very susceptible to rusting, which tends to dull them.

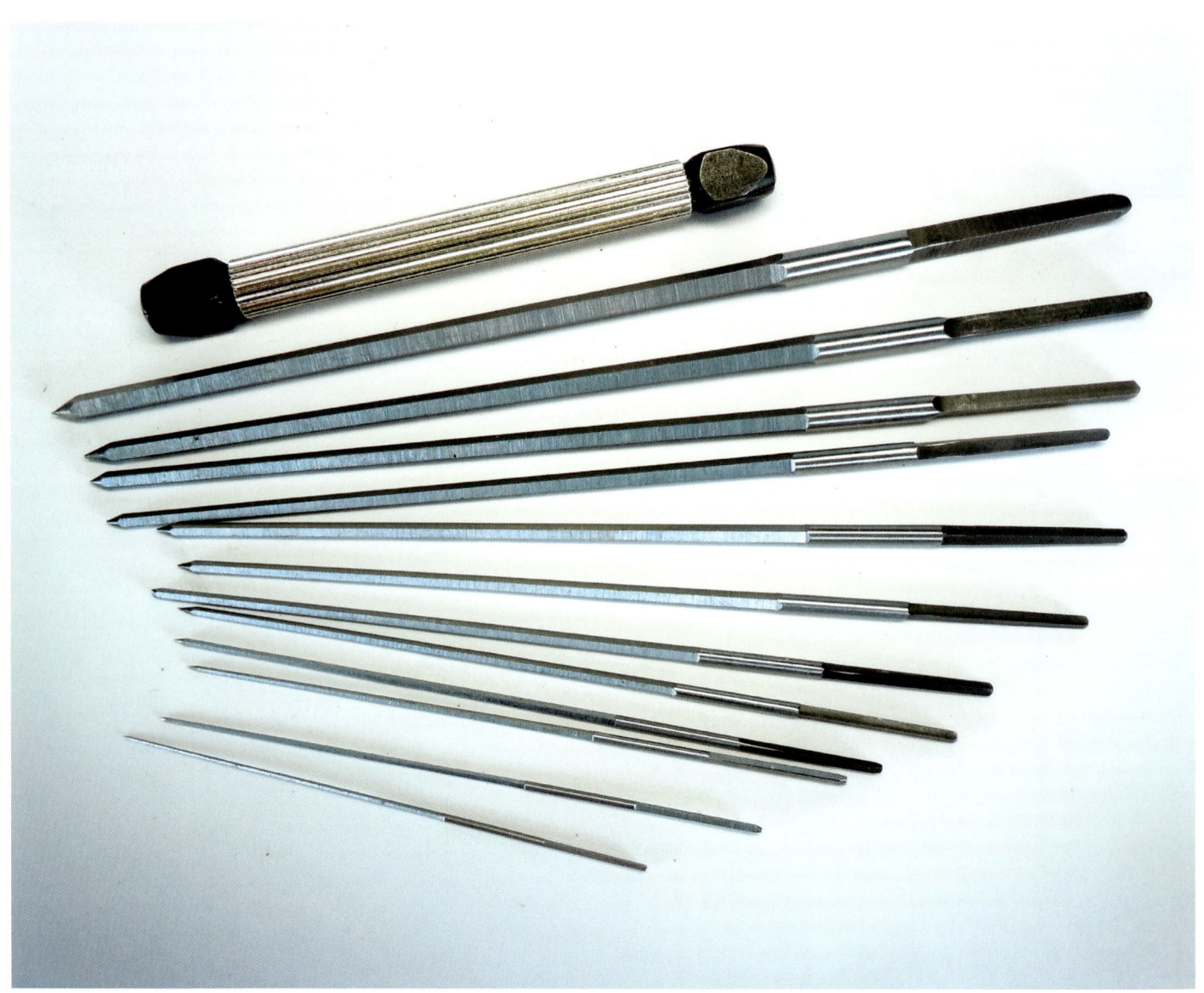

Special Tools: As mentioned earlier, there are those nice-to-have tools. Following is a description of tools I have either used personally or have seen others use. The real issue with any workshop is: "Can one ever have too many tools?"

Bending and Forming: There are tube formers, bending brakes, bending and forming dies, and hole punches. The true test of need is: What is trying to be accomplished versus one's budget. I generally have found that most of my needs are satisfied with a good variety of pliers and a vise. This is also a question of scale as well. Pictured below are tube benders that help keep the tube from collapsing by exerting pressure on the outside of the tube during bending. For very tight bends, it may be better to use rod instead of tube.

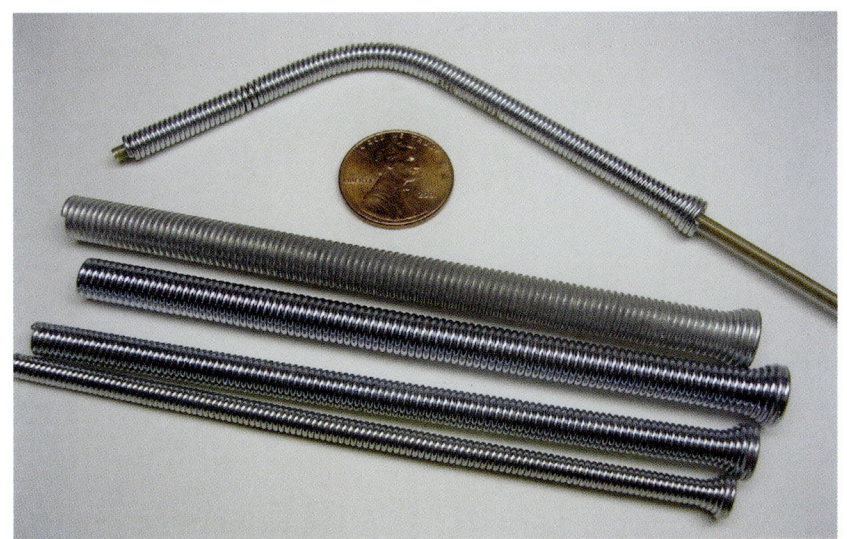

Hole Punches: A hole punch set is very useful; while not cheap, the punches are the best way to make perfect disks and holes up to .75" in thin sheet stock. For best results, cut holes or disks in the standard half hard sheet stock for nice, clean, crisp disks. It is not recommended to use hole punches on soft or annealed sheet stock.

Rotary Tool: A rotary tool with the right mix of accessories for grinding, sanding, and polishing is good to have. With a drill chuck, the rotary tool can also be used like a tiny lathe to file parts such as turnbuckles. Over time, one will accumulate assorted bits, steel cutters, and micro-cutters. I have my dentist save his old used tool bits that are still more than sharp enough for modeling.

When using any kind of power tool whether hand held or bench top, always wear clean safety glasses.

Small Drill Press: A small drill press is a must have machine for the serious modeler. A bench top model with variable speed is enough to handle most requirements for model making. Obtain a drill chuck that goes to zero or a pin vise that can be used in the drill press. Another nice-to-have tool is a drill press vise; there are several models available. I would recommend a small machinist's vise that has V grooves cut vertically and horizontally in the vise faces to hold tubing and rod.

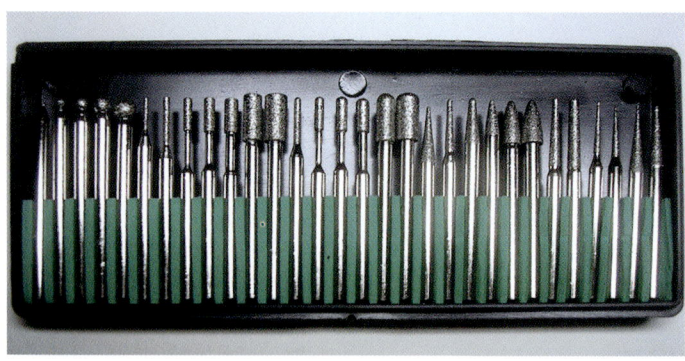
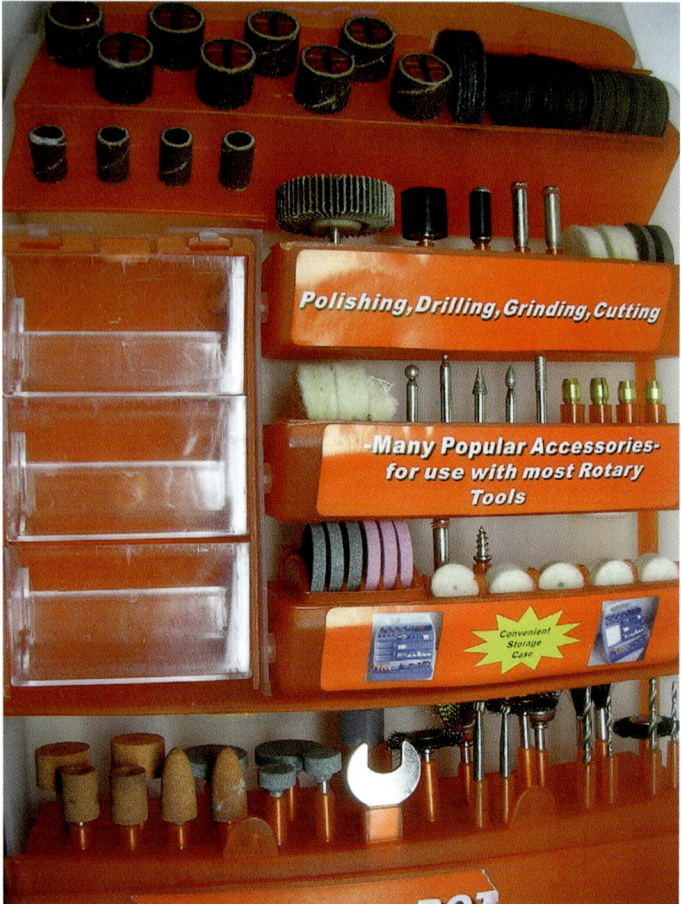

Mini Lathe: A mini lathe is a must have machine for the serious to advanced modeler. Several commercial models are available; again, the bench top size will handle most model making needs. One's pocketbook and skill level will determine the model and number of accessories. The smallest I would recommend is a 7" x 10" size. When purchasing tool holders for tabletop lathes, buy the .125" square size tool holders and cutting tools; the larger sizes will be too large for most operations. One can spend as much or more on accessories for the lathe than the initial cost of the lathe. Start with the basics, .125" tool holders, HSS set of tool cutters, a cut off tool and holder, tailstock drill chuck, and live center. Purchase additional accessories as needed for specific tasks.

Mini Milling Machine: Same comments as for the mini lathe. When purchasing cutters, I suggest two flute for working brass and aluminum. On both machines, read the instructions, follow safety procedures, and learn the limits of the machines and materials you are working with.

A company I recommend evaluating for tabletop equipment is Sherline, Inc. They offer a one stop shop approach to both machines and accessories that will satisfy your model building needs. The challenge with some of the other brands of equipment is the range of accessories available beyond just the basics. Pictured below is just a sampling of their product line offering. Also, make sure to evaluate their bundled machine packages. Do your homework on which machines are best suited for your specific needs.

In ranking tabletop equipment in order of acquisition, I would first suggest the drill press, then the lathe, and thirdly the milling machine. This ranking enables the inexperienced in machining to gradually step up in their learning experience and skills. The internet is also a great resource to explore and learn about tabletop machining. One word of caution is to always wear safety glasses when working with any kind of power equipment; eyes are too valuable not to protect.

Other tabletop machines that are nice to have but not essential are a jig saw or band saw with an assortment of blades to cut wood and metal. I prefer a jig saw because of the tighter cuts it can make on small parts. The best technique for cutting thin brass sheet on a jig saw is to glue the sheet or sheets to a sacrifice wood carrier of .25" plywood. Doing this will give you much cleaner edge cuts and eliminate any edge distortion that would otherwise be caused by the blade on the unsupported edge. A feature to look for is a variable speed setting and a deep enough throat to allow rotation of the material being cut.

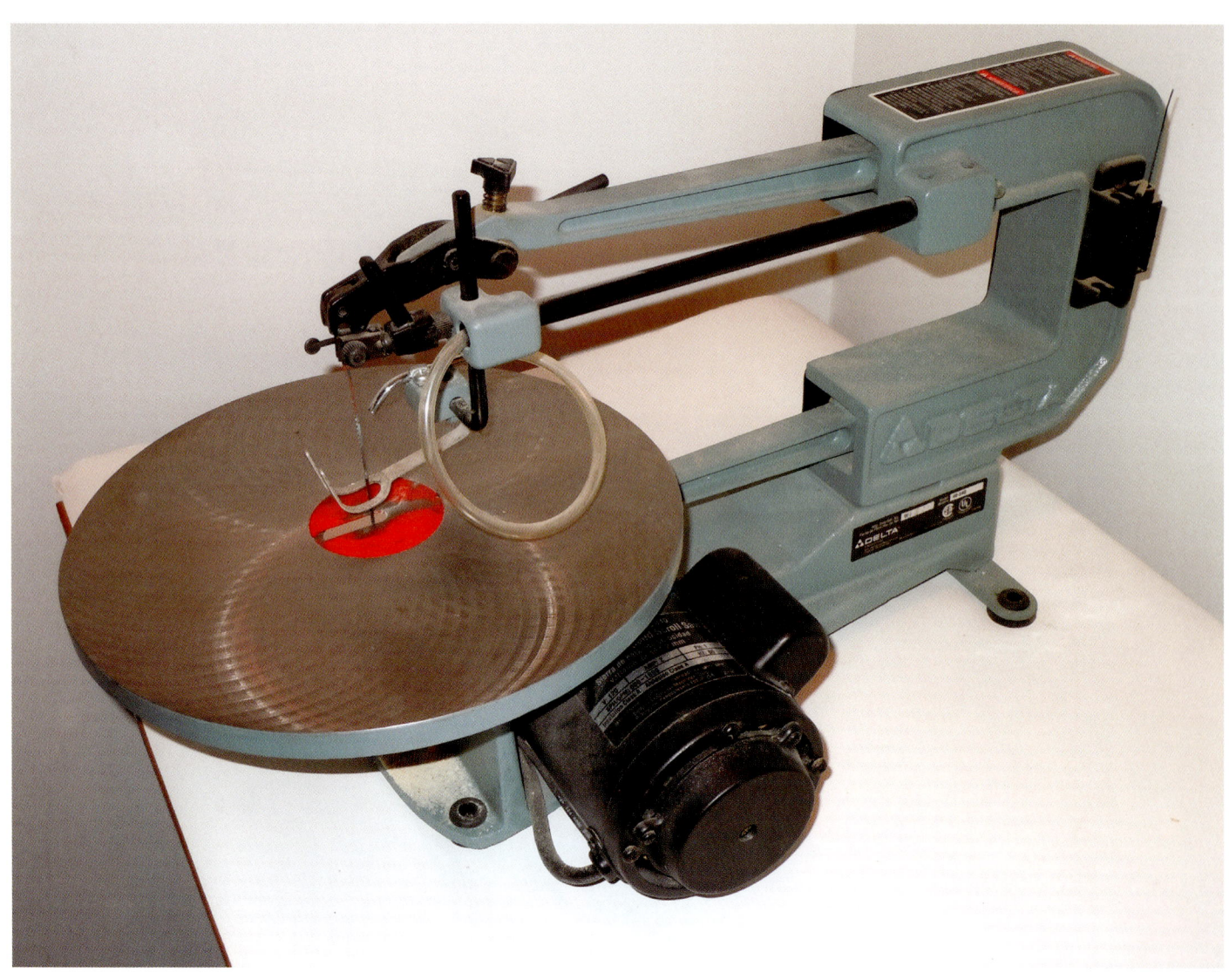

Chapter 3
Working with Brass

Brass, like any other material, has its limits and benefits. One will only learn the real limits by working with the materials themselves. In modeling, one is generally trying to replicate an existing machined shape of a component part. This effort is made easier with brass since there is such a diversity of shapes and sizes available. Many of the standard shapes are assembled in various combinations to form the model or component part. Skill must also be acquired using the various tools; they have limits as well. You can only acquire the skill through practice and in short order learn the "feel" of the tools and brass.

Round Tubing

Round tubing is one of the most-used shapes in modeling. Tubes are used in making fuselage structures, landing gear, and car frames. Tubes must first be cut to a given length for use. There are several ways to cut tubing; the cutting technique used is generally dependent upon the wall thickness and diameter. For small tubes under .125" in diameter, the fastest is using a #11 X-Acto® blade and rolling back and forth on the tube. This technique was explained in Chapter 2. Every tube cut this way will have a burr raised on the tube. It is a good practice to file off this burr that occurs on both the outside and inside. The inside burr on a tube can easily be removed with tip of a #11 blade. If doing several that need to be the same length check their dimensions and file square and to length to match.

To cut larger diameter tubes, a jeweler's saw or lathe can be used. When using a jeweler's saw mark the cut line with a pencil or wrap a piece of masking tape around the tube as a reference line while sawing. Placing a tube in a V block notch will help hold the tube in place while cutting. Use beeswax (or an old wax candle) on the blade. This lubricates the blade while cutting and helps reduce brass buildup in the blade teeth. A good alternative to a steel V block holding fixture for tube cutting can be made from wood baseboard trim that has a detail cut in them. Check with a local lumber supplier, some even give free samples. When using the jeweler's saw, let the blade do the cutting. Extra force will only snap the blade. On larger diameters, rotate the tube and cut around the perimeter. Clean off the burrs and file square to true up the cut ends.

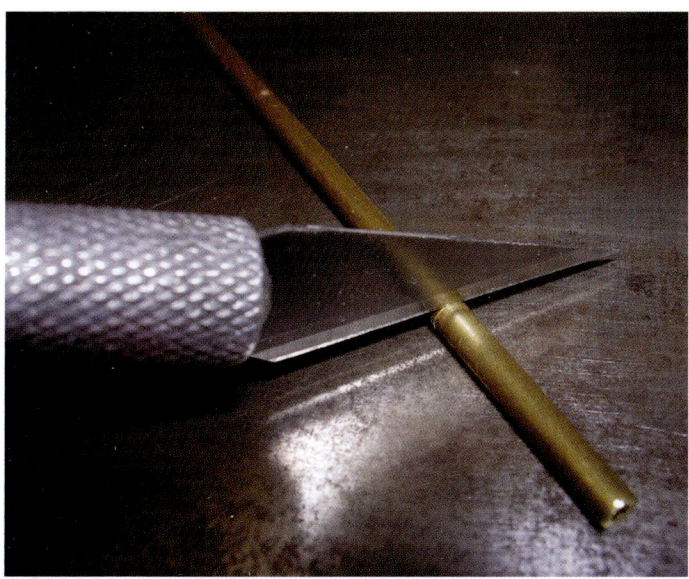

When using a lathe, I recommend grinding a cutoff tool down to about a .032" in thickness and use cutting fluid one drop at a time, if necessary. Again, clean the burr off the inside of the tube. The lathe is especially great for cutting multiple large diameter, thin rings. With practice, experience will be gained to understand the best technique to use based upon the equipment on hand.

Hand operated tube cutters are commercially available and are good for tubing generally larger than .025" diameter. They are scaled down versions of the kind used by plumbers and pipe fitters. They are tightened and rotated and tightened again until the cut is complete. Several sizes are available, just check on limits of tube cutting sizes before purchasing. Thin rings cannot be cut with this type of cutter because of the sizes of the roller guides on each side of the cutting blade.

Now that we know how to cut some tubes, we need to use them. They can be joined to another shape, bent or shaped. To solder tubes together, they **must** have as tight as possible mating fit to have a good solid joint. Remember solder itself has little to no structural strength. So when joining one tube to another one, the end must be filed using a small round file to file a fish mouth joint. This is required whether it is a 90-degree butt joint or an angled joint. Again, the tighter the mating surfaces, the stronger the joint! Fish mouth joints on the end of the tube are easily and quickly accomplished. Start by notching with a small triangular file on one side of the end of the tube on center and lay over to the other side. Once a small "V" notch is filed in as a guide, then use a round file and complete the rounded fish mouth at the required angle. Constantly check the fit for the correct angle, depth, and tightness. See the picture of the airplane fuselage structure. All tubes have been fitted using fish mouth joints. Flush fish mouth joints are best when both ends are soldered to another tube. Tubes can also be penetrated by smaller diameter tubes or rod and soldered. Penetration can be completely through the tube or just one sidewall. A penetration joint will be much stronger than a surface mating joint due to the mechanical interlock. In some situations where strength is required, a penetration joint in conjunction with a fish mouth joint can be made by inserting a smaller tube or rod into a hole drilled in a tube, then the fish mouth tube piece is slipped over the short tube and both soldered in place. Typical joint applications are pictured below. You will see the advantage of building with brass and its strength in small components, especially in the functioning tail wheel assembly pictured below, which is steerable with a working shock absorber. No other modeling material could withstand the weight of the finished model and the stresses of functioning.

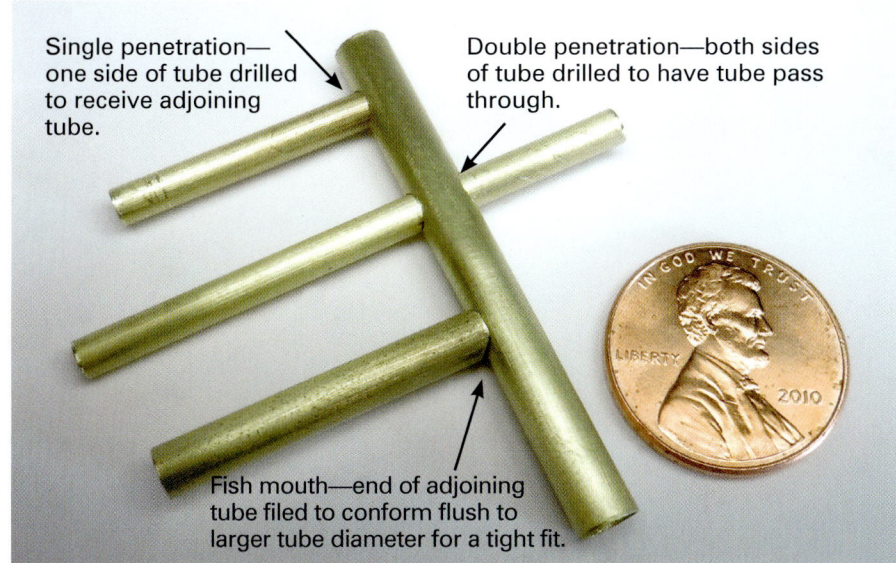

Typical joints for tubes and rods.

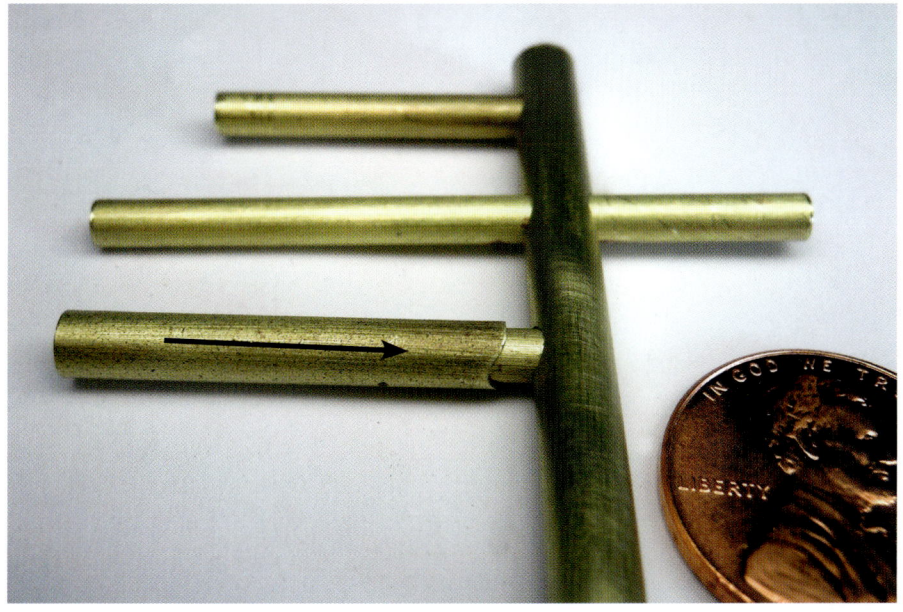

Fish mouth used with single penetration short tube for maximum strength joint. Slip tube to be flush with adjoining tube and solder both at the same time.

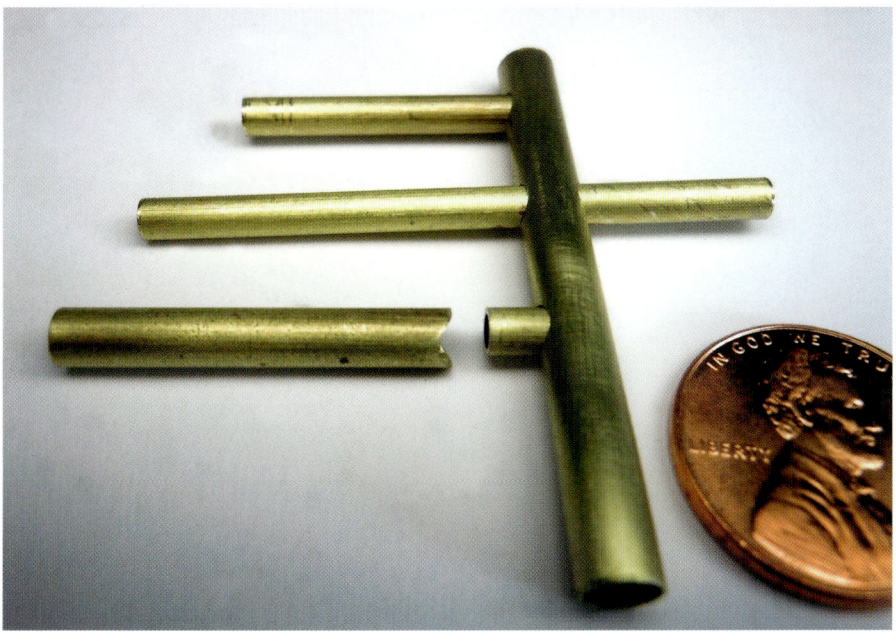

Strongest joint for joining tubes. Index fish mouth onto a short penetration tube for mechanical interlock rather than just a single butting surface.

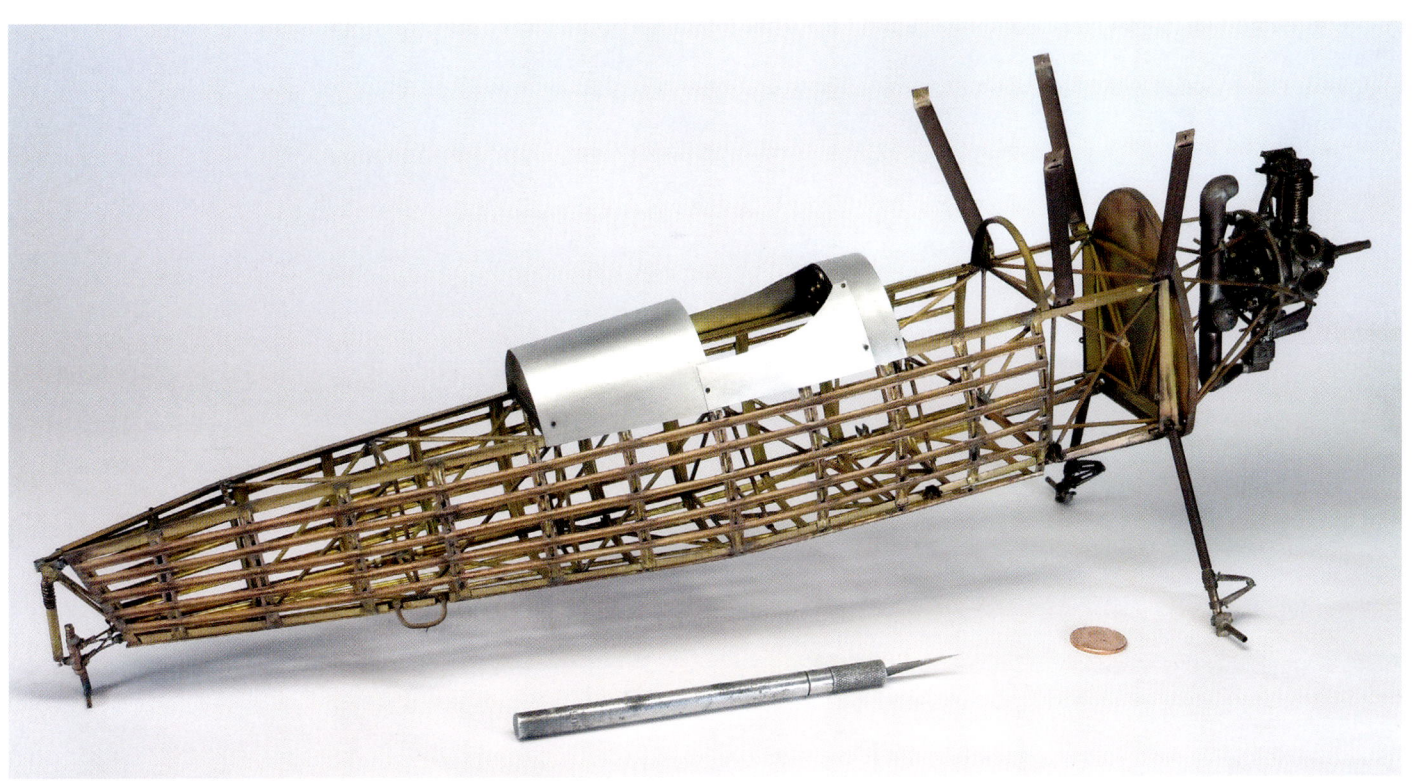

Tubes can be joined to rectangles and flat sheet the same way. They can be surface mounted and soldered or the stock can be drilled and the tube inserted in or through the hole, then soldered. Again, tightness of the hole is best for strength. Surface soldering tube to .005" or .010" sheet is a great way to make instrument bezels and gauges. Use a length of tube to solder the sheet onto, trim excess sheet away with scissors using the end of the tube as a guide, file clean around joint and cut to required depth of bezel or instrument gauge. Add trim rings or flanges as necessary. This is a great technique to make instruments for cars, boats or planes in any scale.

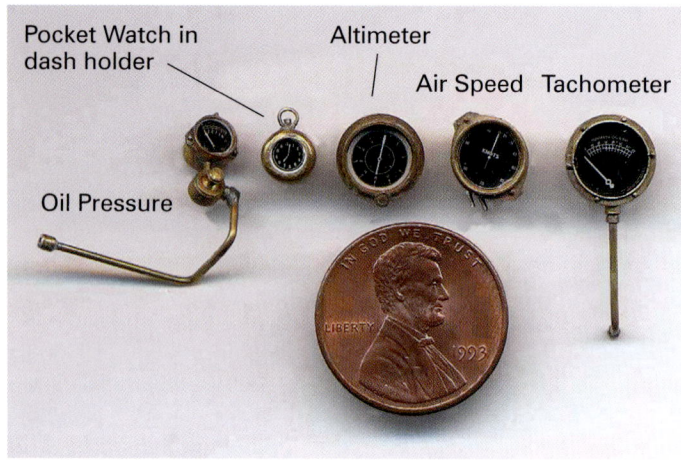

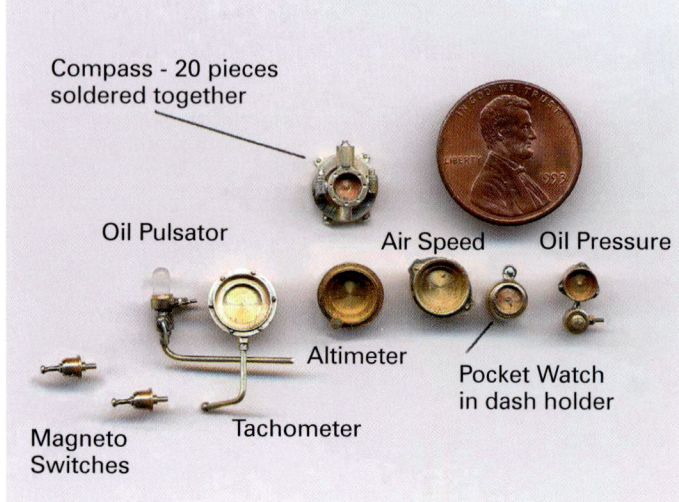

Sopwith Dashboard Instrument Bezels.

Bending a brass tube can be accomplished in several ways. Tightness of bend is dictated by wall thickness and diameter. If a very tight bend is required for a feature, solid rod must be used. To bend small diameter tube, .187" and less, bending tools are commercially available that help reduce the tube walls from collapsing. The challenge in using them is bending complex shapes. Making more than one bend in different directions on the same tube is difficult. In working with .125" or less, the best way is by hand using fingers and **bending slowly** over paper plans and pinning down as the shape is worked. In some cases, T-pins can be used as bending points working around a shape. A reference example for this technique is the forming of airplane rudders, elevators, and ailerons. See previous picture of tail plane assembly. To bend larger shapes or circles, bend slowly around a mandrel. A mandrel can be anything round and solid that can resist the bending force, such as paint jars, tool handles, punches, etc. Keep in mind that the mandrel diameter should be smaller than the desired diameter of the shape or circle because of spring back in the brass tube. Using a mandrel also reduces unsightly kinks in the bend. At times, there will be the need to final fit using thumbs and index fingers to adjust the final shape and flatness. Once the desired shape is accomplished, solder in place as required. This technique can also be used on rod as well. When given a choice, always try to use tube instead of rod for ease of soldering, weight reduction, and strength.

Depending upon the diameter of stock being used, annealing of the tube or rod may be required to get the desired shape. Annealing is the process of "softening" metals. Brass is a non-ferrous metal, so to anneal, heat the stock to a dull red color with a butane or gas torch, then quench quickly in water. This is the opposite of ferrous metals, which are hardened when quenched in water or oil. Sheet stock, when hammered into shapes, will have to be annealed during the forming operation because of work hardening by the hammers. Annealing is one of the processes that will require some experience to get used to.

Finished intake manifold ready for installation.

Profiles, Shapes, and Rods

Working with profiles, shapes, and rods, follow the same basic premises as described in the round tubing section above. Profiles can be rectangles, squares, and hexagons, and are considered tubing. Shapes are formed or extruded stock, such as angles or extrusions like H-beams. Rods are solid and can be round, square, rectangular, and hex. They all can be cut, bent, formed, drilled or shaped as required. Cutting is usually accomplished with the jeweler's saw or very carefully with a fiber cut off wheel in a hand held electric rotary tool. Size generally dictates which tool or technique to use for cutting.

A stock of particular interest for many modelers is the hex rod. Hex rod is available in various dimensions down to .046". These small sizes can be used to simulate bolt heads and nuts as required on engines or other applications. See picture of engine details.

The following is a brief description on how to do both if you do not have a lathe at hand. Start with required size of hex rod. Whenever possible, drill holes the diameter of the hex rod in nut or bolt head location. Cut hex rod to length using hand shear, end cutters or jeweler's saw .062" to .125" in length. After each cut of the rod, file the end of the rod square with file, and then cut next length. Now, insert cut rods into drilled holes with cut end sticking up. Carefully solder all rods in holes. If situation does not allow soldering, then carefully epoxy in place using fifteen minute epoxy. Fifteen minute epoxy is stronger than five minute and will give you more working time. Once parts are soldered or glued in place, with a fine needle file, carefully file down all cut ends sticking up to the same height. You now have bolt heads. To simulate nuts, do the same as bolt heads, but now very carefully drill into the bolt heads with an appropriately sharp drill size just enough to hold a short length of rod to simulate bolt end. Solder or epoxy rods in place and file the ends square between length cuts. Once the glue has cured, file to the same height. This operation can be seen in the picture of a propeller hub where .046" rod was used for nuts and drilled using a .020" drill bit with .020" rod inserted to simulate bolt ends sticking out of the nuts. Larger nuts and bolts can be simulated using 00-90 hex bolts. Insert cut off bolt into hole for a bolt head, invert and solder or glue and cut off to simulate a nut. This can be a pricey option if many nuts and bolts are required. Hex rod can also be turned using a lathe to create simulated nuts and bolts as well. Below is a work in progress photo showing various applications and sizes of nuts, and once all are soldered in place, excess solder will be cleaned off. Yes, soldering does get messy at times, but the second picture of the rotary engine shows the results once it is cleaned up.

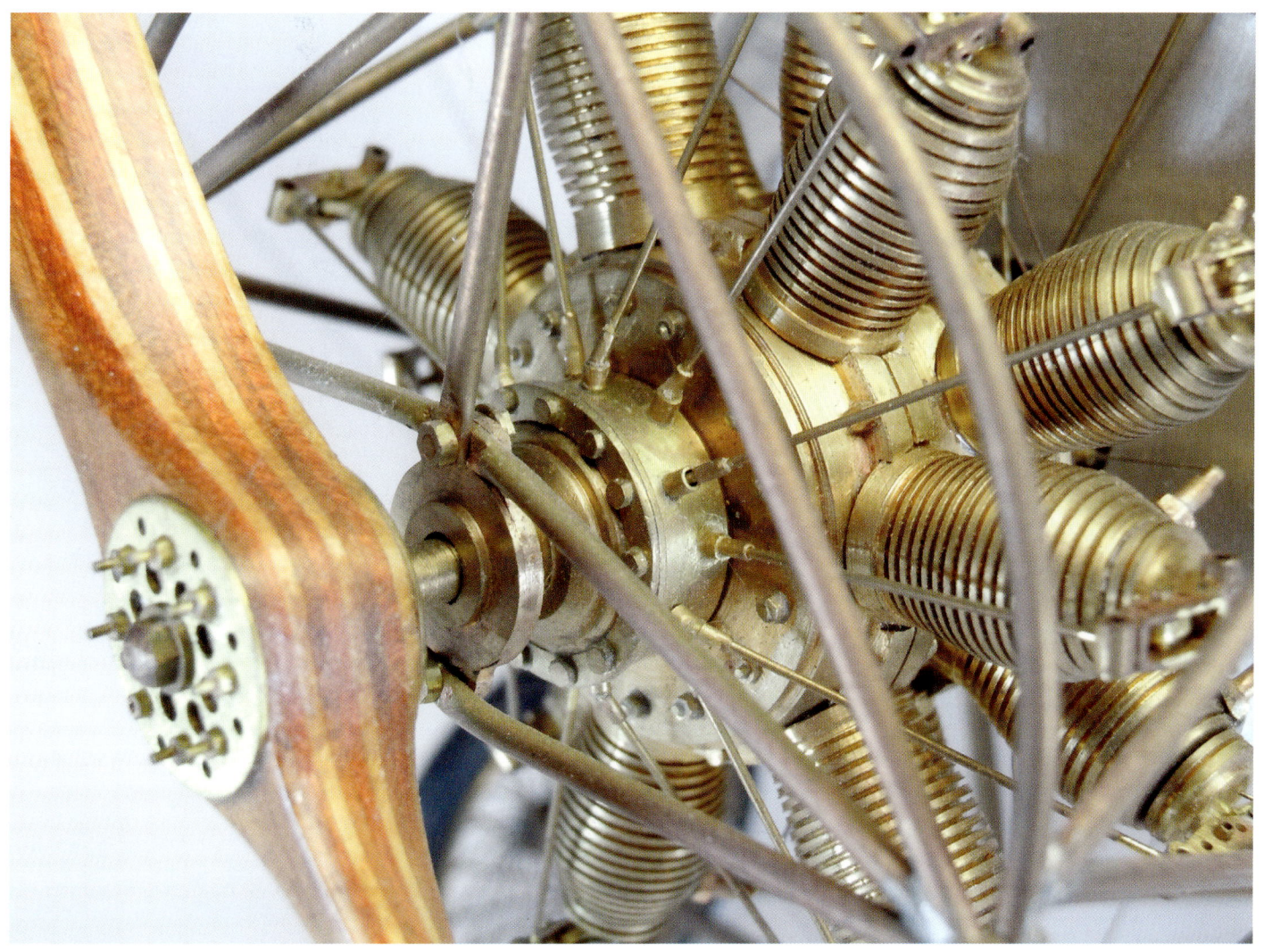

Sheet Stock

Sheet stock for modeling purposes is available at most hobby retail outlets or can be ordered directly from suppliers online. The sheet stocks most commonly used are 4" x 10" sizes in thicknesses, .005", .010" and .015". A word of distinction here about sheet stock versus foils and shim stock: foils are extremely thin, difficult to work with, and generally are applied to a substructure for the appearance of brass; shim stock is used in industry for leveling purposes and is generally made of a hard brass alloy. Some special applications may require the use of these types of sheet stock, but the majority of general purpose modeling needs, such as fuel tanks, oil tanks, and ammunition cans may be formed from the .005", .010" or .015" stock.

Generally, try to use the .005" stock because it is easy to cut with scissors, hand form, and solder. Forming can be accomplished with smooth duckbill pliers for small 90-degree bends to forming cylinders and rounded shapes using round mandrels or wood forms. The complexity of the shape really dictates the technique to be used. In round gas tanks for example, form the cylinder carefully around a round object (mandrel) smaller than the final diameter desired because of spring back. Use a longer length and cross cut to final size once formed into a cylinder with about .125" overlap for the solder joint. Using narrow masking tape, tape to final diameter in the center of the cylinder or shape, leaving ends exposed of tape. Then, flux and tack solder the ends. Remove tape and carefully solder center area without getting too hot toward ends and breaking those joints. Once joined, work the solder joint smooth from the center to one end, then repeat to the other end. An alternative method is to use binding wire, very thin malleable black steel wire, which is wrapped around the shape, and then the ends are carefully twisted to hold the desired shape. Binding wire is especially useful when working close to an already soldered joint that may get hot enough to break loose under heat; the wire "binds" it tightly in place and prevents that joint from breaking. Then, the tank ends can be cut to size, trimmed and press fitted into place. Press fitting is critical to hold in place and space equally around the shape until soldered in place. A word of caution for any hollow form, **always provide a vent hole for hot gases to escape during soldering operation**. In gas tanks, locate and drill the filler neck hole before soldering the second end in place.

To achieve a rolled seam effect, form a .020" wire ring that press fits inside the diameter of the tank cylinder or shape and solder using a minimal amount of solder. Flat strips can be used instead of wire for a different look. Below is an example of a round gas tank in a WWI airplane.

Forming three-dimensional shapes can be accomplished with sheet stock. However, it requires using "soft" brass that is generally a special order, and forming hammers like those used in auto repair over wood or steel forming bucks. Using these requires extreme patience and skill, but extremely complicated forms can be accomplished, such as automobile bodies.

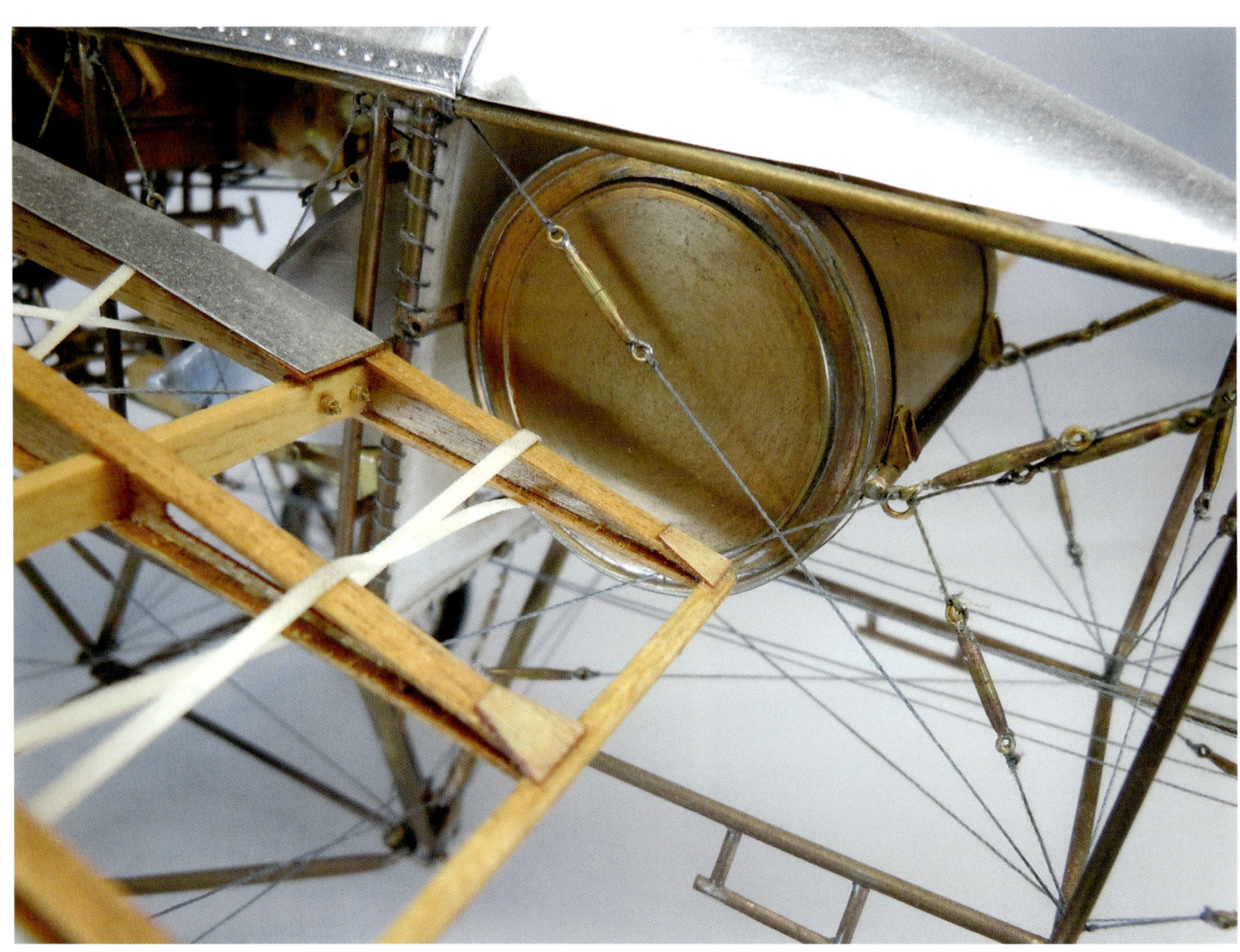

Model Building

Bringing together the various tools and techniques discussed so far can result in just about any model desired, whether car, airplane, boat or armored vehicle. Models are generally comprised of a combination of basic geometric shapes that have been described above or are commercially available. Speed of building is another great benefit of using brass. Once soldered, building goes on without having to wait for glues or adhesives to set. Brass is also forgiving when one makes mistakes —joints can be unsoldered, relocated, and soldered again. Strength of parts and structures are also a great benefit. Brass has a great strength to weight ratio that is difficult to achieve with other modeling materials. In fact, many are amazed at just how strong some structures can get and how much rough handling can occur during construction without causing damage that would normally destroy other materials. Brass can also be easily painted to replicate any finish or color desired. It can also be easily stripped of paint using paint stripper without damage to the part, unlike wood or plastic. Probably the greatest disadvantage is that it requires the use of soldering irons and developing the skills to use them, but soldering, like all skills, is easy to learn but requires practice to perfect.

Soldering

"There is no magic, but there are do's and don'ts."

Soldering is the secret to successful joining and modeling with brass. Soldering is a learned skill. There is no magic; however, you must follow the correct process to attain success. Practice will perfect this skill from very basic joints to very sophisticated joints.

As with any task, having the correct tools for the job is a must. The most expensive mistake that can be made is buying cheap tools! There are many levels of quality in soldering irons and more than one is needed to accomplish various soldering operations. In fact, some operations may even require the use of a small butane torch based upon the mass and size of material to be soldered.

"An expert at anything was once a beginner."
Fortune Cookie.

Soldering Irons

Soldering irons are sold in terms of wattage, ranging from 25 watts on up. For the purpose of modeling, most operations and solder joints can be accomplished with 40 watts, 60 watts, and 120 watts irons. There are many models and styles to choose from, so when purchasing consider the following points. Features to look for are a straight, easy to hold handle, at least a five-foot electrical cord, and some models have a light to indicate that the iron is on. Look at how the tips are attached. Some have set screws, some screw in. Tips should be considered as disposable items, so tip replacement is critical. The screw in type generally tends to break off after extensive use because of flux (acid) spattering and heat cycling that corrode the threads and shear off when trying to replace the tip. Once sheared off, the iron will require drilling out the broken end and tapping new threads – a pain to do but it can save an iron. This obstacle can be overcome by making it a practice to unscrew the tip from time to time and clean the threads with a steel brush. The set screw types require the same care on the set screw, again because of flux spatter.

The soldering iron tips come in different shapes for different applications. For 40, 60, and 120 watts irons, purchase round pointed and chisel style tips. The round pointed type is good for getting into tight areas of tube joints, while the chisel point is good for flat sheet work like soldering joints on gas tanks. The 120 watts chisel point is also useful in unsoldering joints when a mistake is made or rework is necessary.

Over time, with the use of the corrosive flux and heat cycling, the tips will break down, developing holes and craters. These can be filed away using a course file to renew the tip face or shape. It is always a good idea to purchase extra tips to have on hand when needed. To preserve tips, it is a good practice to keep a wet sponge on a ceramic saucer handy to wipe and clean the tip often during soldering operations. The use of a "tinning block" (sal-ammoniac) helps renew a blackened tip. Clean the tips of your soldering irons after each session of soldering.

Solder

I have used many solders over time and to date the most versatile for general purpose construction is a product called StayBrite®, which is a lead-free silver content solder that comes with its own flux when purchased through catalogs or hobby stores. The advantages of this type of solder are its strength and ease of use near other joints without binding wire, which saves time. It also cleans up well with small files with less buildup in teeth than plumbing type roll solders available in hardware stores. Files get expensive after awhile. Remember the comment about cheap tools; it also applies to materials. A roll of StayBrite® goes a long way, depending on the scale of model being built. The cost effective way to purchase this brand of solder is from a local industrial plumbing supplier in one-pound rolls. If they don't have it in stock, they can order it for you. I generally use the .032" diameter solder, but it is available in other diameters. Purchase a pint of flux at the same time.

Fluxes

Use the type of flux best suited for the type of solder being used. Fluxes come in both liquid and paste form. Both are best used sparingly and applied with a cheap, small, round artist type paintbrush. Fluxes chemically clean the metal surfaces to be joined. Off gases from fluxes during soldering operations can be nasty and cause headaches over long use. Use only in well-ventilated areas, read instructions, follow directions, and take fresh air breaks if working for long periods of time. Clean the areas around the joint to remove flux splatters when soldering operations are finished. Use acetone on a Q-tip® or paint brush and wipe dry. Some fluxes are water soluble, so check before cleaning. Splatters, if left and not cleaned, will continue to etch metal surfaces over time. Remember, they are acid-based.

Soldering Operations

The most important factors to keep in mind when soldering are tight fitting joints for strength, high enough soldering iron temperature for solder to liquefy and flow **(remember solder will always flow to heat)**, clean surfaces using the proper flux, and the correct type of solder for the job at hand.

Solder joints are many and varied from sheet to sheet, tube to tube, and various other combinations. Sheet to sheet may include sheet stock of the same or different thicknesses. When soldering, always concentrate the heat on the thicker wall stock and the thinner will be heated as well. Clamp or hold together the parts to be soldered. A tip to keep in mind is to "tin" the thicker part(s) before adding the smaller ones. Tinning is adding a thin film of solder to a mating surface to facilitate solder flow. Keep in mind that solder may cause parts to "float" if not clamped when the solder is in its liquid state, much like hydroplaning of car tires in heavy rain. Sometimes assemblies cannot be easily clamped or held together due to size; then, a mechanical interface must be used, such as pinning with small rods inserted into drilled holes to hold and register parts until soldered.

The accompanying picture shows the building of a Mercedes engine block using various layers of brass stock rough cut to both inside and outside shapes to help reduce the overall weight. The top picture shows the outside view of the layers clamped together using a metal spring clamp and machine screw to hold layers firmly in place while soldered using a propane torch and Staybrite® solder. Note the cylinder and bolt hole locations were drilled prior to the soldering operation. This is an example of planning ahead to drill the holes while the sheet was double taped to a piece of hard wood, enabling clean drilled holes. When working with layers, it is critical to have smooth, flat mating surfaces that are fluxed before clamping together; this will reduce the visibility of the soldered joints once finished. The advantage of this layering technique is that complicated parts and shapes can be fabricated without fairly sophisticated machining techniques and milling machines.

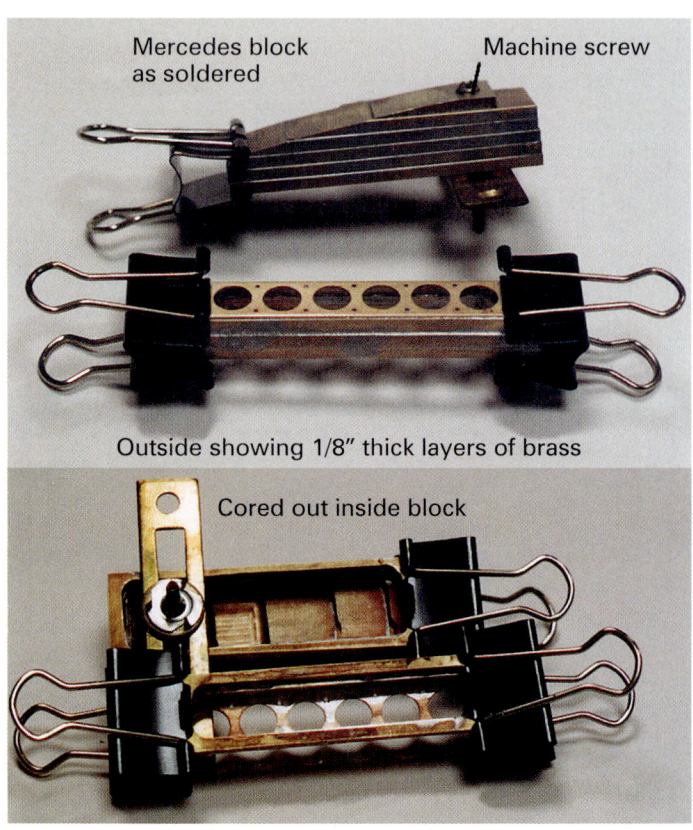

Here is what the finished engine block filed to shape looked like when completed. This was all shaped using only files by hand, with no machining involved. Once the final shape was finished, two flanges were cut and soldered to each mating half.

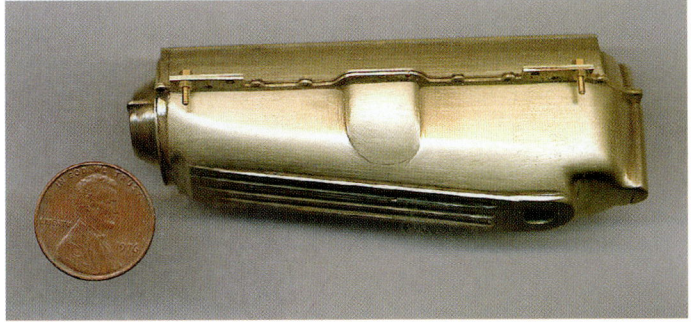

Here is another example of using layering to build up a part without the use of machining to accomplish a difficult part like the head of a Model T engine. The lower part shows the relief required for the valves along with the other bolt hole and spark plug locations. On the top part, the larger bolt shoulder tube holes were located prior to soldering. The middle layer spark plug holes were tapped with the needed thread size.

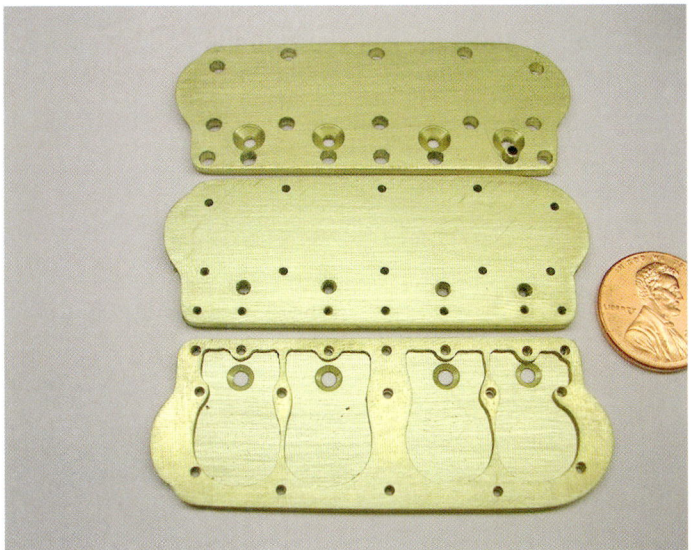

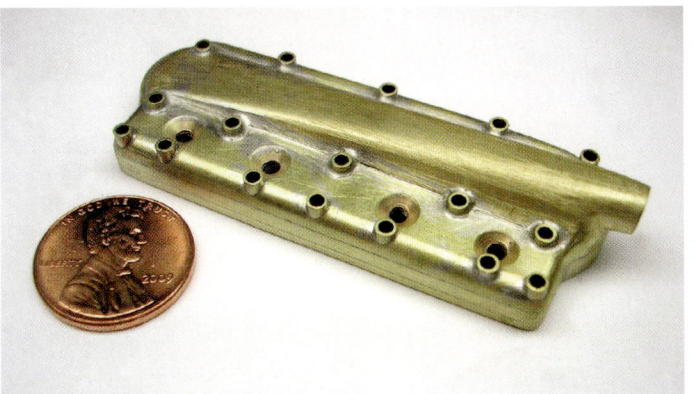

Tube to tube, and tube to rod, and tube to sheet are other possible combinations to be soldered. A common tube joint is a butt joint of the same diameter tube, which requires a smaller tube slip fitted into each longer tube to be joined. In this case, the hot solder is added at the seam to flow around the joint and in each direction on the short inserted tube as pictured below.

Tube to tube joints can get more complicated based on the part geometry trying to be modeled. Whether tubes are the same or different diameters, a "fish mouth" joint may be required. The fish mouth is the filing of the end of the tube with a round file to conform to the tube being soldered to at the desired angle. An alternate method, if a small diameter is to be soldered to a larger diameter tube, is to drill a hole the same diameter as the tube in the larger tube, then insert the smaller tube in the hole and solder as needed. These operations were explained earlier in this chapter. This applies to sheet stock as well.

The same principles apply for tube to rod joints only with the heat being concentrated on the rod, which has more material to be heated for the solder to flow. The concept is to always use the thicker part as the heat sink for a strong joint.

Rod to rod joints are in principle the same as tube joints but usually requiring more heat to solder together due to solid material and more mass.

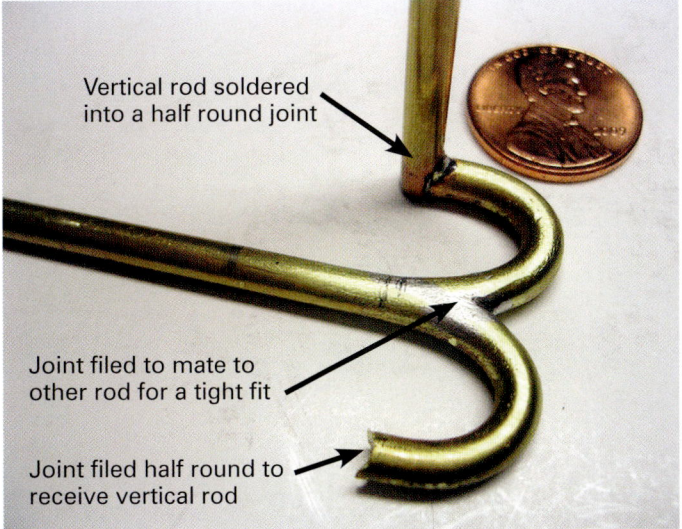

Here are subsequent pieces added to the above manifold. This is when a drill press comes in handy to drill out holes and locate pins and rods.

Retaining collars soldered in place.

1/16" rod used for locating carburetor.

Finished intake manifold ready for installation.

Here are subsequent pieces added to the above manifold. This is when a drill press comes in handy to drill out holes and locate pins and rods.

In the accompanying picture are examples of tube in tube in conjunction with sheet stock to form the core on an engine block. Looking at the bottom, you can see where the cylinder wall tubes are press fitted into the lower sheet as well as the top sheet while passing through the middle sheet. Then, the smaller tubes comprised of three different sizes form the locations for the push rods. In this case, the smallest diameter press fits into the top and bottom sheet stock while passing through the middle layer with the second size tube indexing in the middle layer and a second set of the same size resting on the lower level. These form the spacing gap where the springs will be located and sawed off at the shoulders. The third size of tube is slipped over the second size, forming an equal spacing shoulder and soldering surface for the middle and top sheet layers. Hopefully, the picture helps this description, but it also demonstrates how thinking

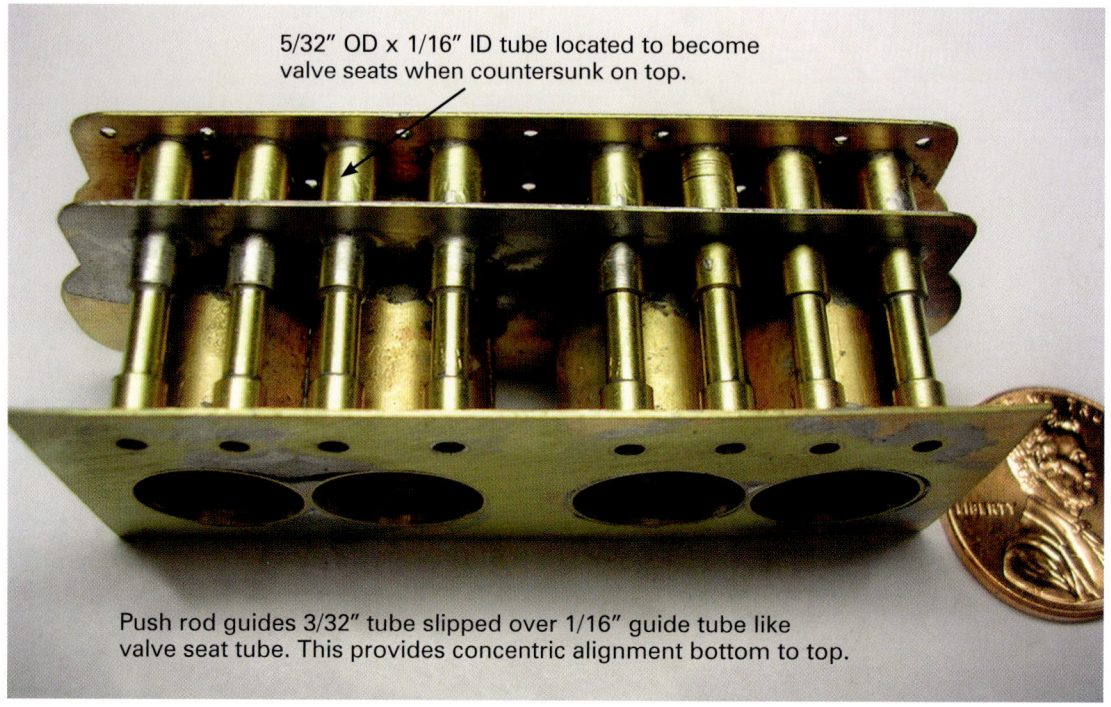

5/32" OD x 1/16" ID tube located to become valve seats when countersunk on top.

Push rod guides 3/32" tube slipped over 1/16" guide tube like valve seat tube. This provides concentric alignment bottom to top.

Rabbit hole joints provide a technique to join parts where it may not be possible to solder the end of parts being joined. This concept relies on capillary action to flow the solder down tight fits to get the solder where it needs to be. An example is the drilling of a hole in a tube, then inserting a smaller tube that has been fluxed. The melted liquid solder is introduced into the drilled hole, and it will flow down the inside of the two tube walls. See example pictured here of how this technique is used on a more complicated joint using sheet stock.

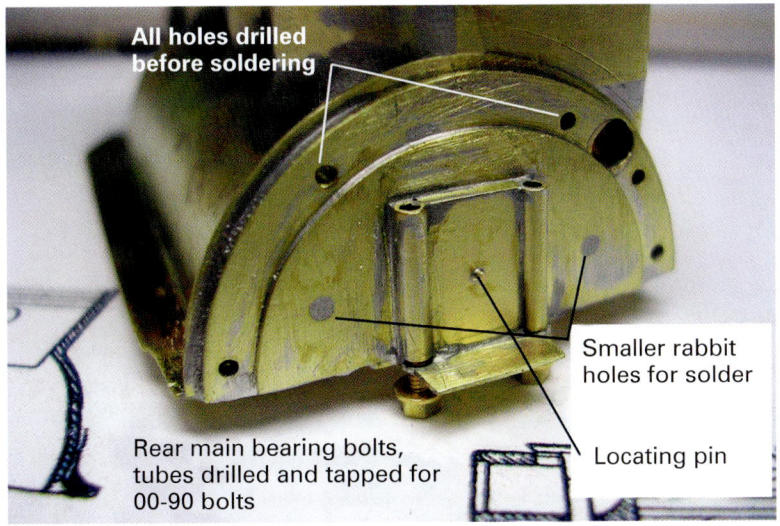

All holes drilled before soldering

Smaller rabbit holes for solder

Rear main bearing bolts, tubes drilled and tapped for 00-90 bolts

Locating pin

In the associated picture, it would have been impossible to get solder to flow and solder the entire mating surfaces of the three half round pieces. Two holes were drilled into the middle part to allow solder to be introduced and reduce the distance the solder had to flow to provide a strong joint and keep the perimeter edge clean. When soldering, watch the edges to see when the solder reaches it and move the hot iron across the surface, and the solder will flow to the heat assuring a complete solder joint. Subsequently, two more holes were drilled in the third part, hiding the middle part's rabbit holes and were filled and filed flush when the soldering operation was completed. The real benefit here is being able to fabricate the part in layers of sheet stock and eliminate the need for sophisticated machining operations.

Sometimes, a shape or form is needed that is non-structural. One way to accomplish this is with solder filling. This is the same principle used in the early days of automobile building and hot rods using lead to fill and blend body joints. This is a valuable technique if the part is to be electroplated later on. Below are pictures showing various examples of filling operations. In the case of the engine block, a large radius was needed to form the transition from the cylinder wall to the block side wall. Here were cut and fitted small triangle pieces that fit as tight as possible between the two adjoining cylinder walls, then filled with 50/50 solder. When doing this, the iron must be very hot and, by holding the block horizontal, it was able to be rotated to get the molten solder to flow as needed. Once cooled, it was then filed, shaped, and cleaned. A second and third application of solder was repeated to form the desired radii per the original block. If this part were to be painted, a surface glazing putty could have been used; however, since the part was to be nickel plated, it had to be completely formed with solder. Keep in mind, in filling operations, it is better to use 50/50 soft solder since it melts and flows and at lower temperature than the Staybrite® solder.

The picture below shows the initial filling operation and how rough it can look. This was followed up by a second filling operation. Looking at the right two cylinders, you can see the top point of the triangular filler piece. Solder filling like this may require several steps holding the part at different angles to get the liquid solder to flow as you wish. This is especially true on larger parts like this engine block.

Here are subsequent pieces added to the above manifold. This is when a drill press comes in handy to drill out holes and locate pins and rods.

In the accompanying picture are examples of tube in tube in conjunction with sheet stock to form the core on an engine block. Looking at the bottom, you can see where the cylinder wall tubes are press fitted into the lower sheet as well as the top sheet while passing through the middle sheet. Then, the smaller tubes comprised of three different sizes form the locations for the push rods. In this case, the smallest diameter press fits into the top and bottom sheet stock while passing through the middle layer with the second size tube indexing in the middle layer and a second set of the same size resting on the lower level. These form the spacing gap where the springs will be located and sawed off at the shoulders. The third size of tube is slipped over the second size, forming an equal spacing shoulder and soldering surface for the middle and top sheet layers. Hopefully, the picture helps this description, but it also demonstrates how thinking ahead and planning subsequent operations is not only valuable but also time saving. This would be a very difficult part configuration to machine or cast but is fairly quickly accomplished by fabrication.

First, solder fill the parts with 50/50 soft solder, which is a lower melting point solder. The key to solder filling is a very hot soldering iron and holding the part with leather work gloves or pliers because of the heat and rotate the part as the hot liquid solder is applied off the chisel tip of the iron. Liquid solder will flow as the part is rotated due to gravity.

The picture below shows the final clean up and blending using various tools. The most valuable is the back edge of a #11 X-ACTO® blade for scraping away and shaping. The final blending was carefully done by wet sanding with 320, then 400 grit sandpaper.

This is the end that the radius was formed using small duckbill pliers. Look closely and you can see the butt joined seam just next to the shadow line.

The finished sanded block ready for copper plating.

The associated picture shows the finished block copper plated. The copper plating does two things: first, it acts as a binding layer for the nickel plate. However, the more important function is to provide a first look at the finished part. At this point, you can further finish the block by additional soldering or sanding and plate again with copper until satisfied. This is of particular value blending the transitional edge from solder to brass using the copper plating as a miniscule filler to blend. Below is the nickel plated block.

Here is another example of a solder filled formed part. First, the oval tube was cut and press fitted into the flat plate at the required angle, and this joint was silver soldered. Then, brass filler pieces were added to reinforce the base connection, and then soft solder was used to fill completely around the part. The part was held with needle nose pliers and rotated as puddles of liquid solder were added and allowed to chill before rotating again. Once enough solder was added, the radius was filed, shaped as needed, holes drilled, and finish rings added. Below is the shaped radius with top rings in place.

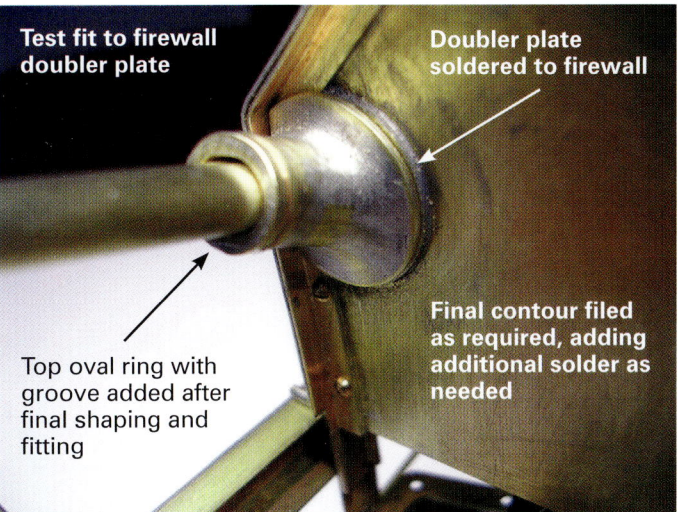

Here is the finished bracket ready for test fitting. Oval tubes are easily formed by gently tapping a round tube laying flat on a steel block with a hammer.

Finished rough solder build up of bracket. Plate was filed so oval tube indexed into it at the correct angle and soldered. The two additional tubes and a bottom side plate were soldered in place to build up thickness. Then solder was flowed to develop radius.

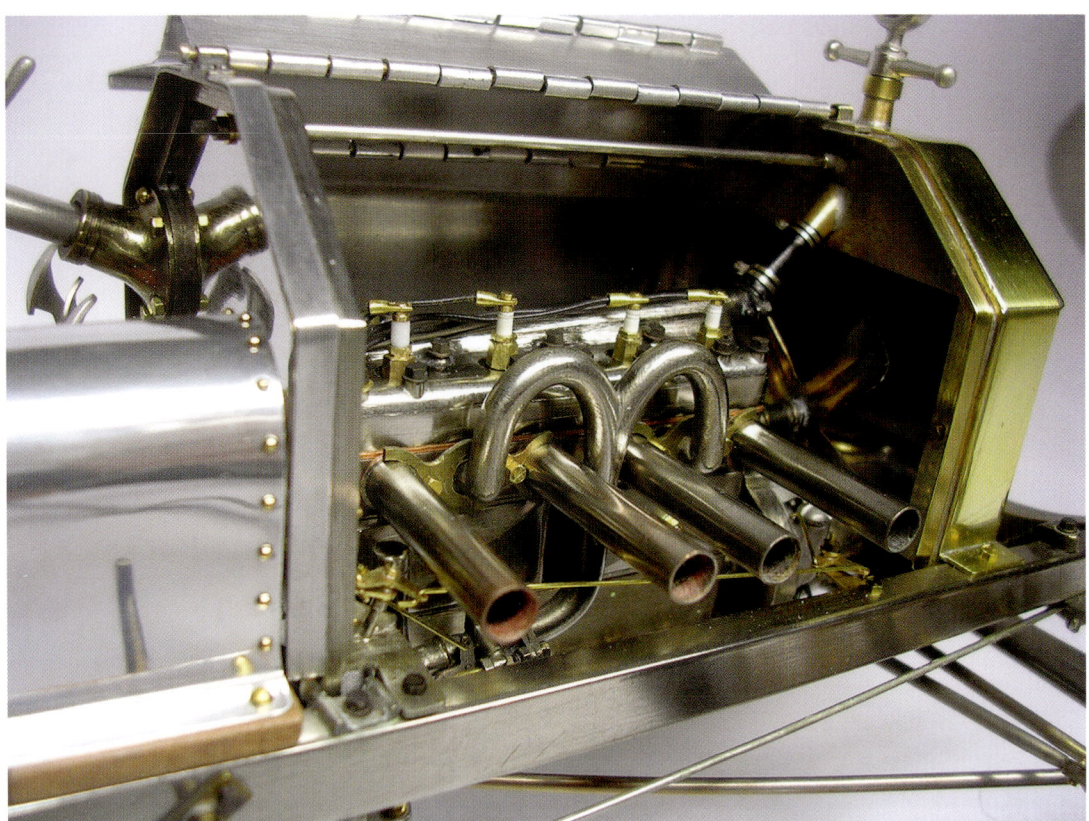

Here is the finished part by itself ready for plating after being test fitted to the firewall with the steering column in place. Finishing parts for plating requires much more patience and time than parts that are to be painted. Many of the finishing principles are the same as other building techniques once the parts are fabricated. The picture below is the finished unit brass plated per the original finish. Notice how the solder filler has a clean, smooth blend to the final shape once plated. Brass is the only modeling material that enables the quick building of a part like this.

Chapter 4
Part Fabrication Using Sheet Stock

Having covered some of the fundamental brass building basics, let us look at some actual part fabrication using various techniques and methods. A common challenge of model building is working with sheet metal or flat stock. Brass sheet that is available at retail outlets is what is referred to as a half hard alloy. Soft or malleable brass must be specifically asked for and is used for forming shapes with hammers or other tools. Half hard sheet can be made slightly malleable by heating with a propane flame and quenching in water. It is best to first cut the needed shape or pattern, then heating it rather than trying to soften an entire sheet.

The following section will show the building of the engine block for a Model T racer that has some interesting and challenging contours using .015" sheet stock. Since this engine will have working pistons and valve train, driven by the hand crank, the crankcase must be fabricated as a hollow unit to achieve the necessary clearances. Other thicknesses that are used are noted for particular parts that appear in the pictures. The following fabrication principles can be used and applied to achieve various part configurations. These parts will eventually be nickel plated, so the finish must be very smooth. Parts to be painted need not be finished to the same high level.

A printed paper pattern was spray glued to two sheets of .015" brass sheet that also have been spray glued together. In this way, two identical shapes were cut out and holes were drilled in identical locations. In this case, the pieces were drilled with #61 drill bit to enable 00-90 threads to be tapped in the lower layer and clearance holes in the upper piece. The two sheets were separated with lacquer thinner to breakdown the spray glue.

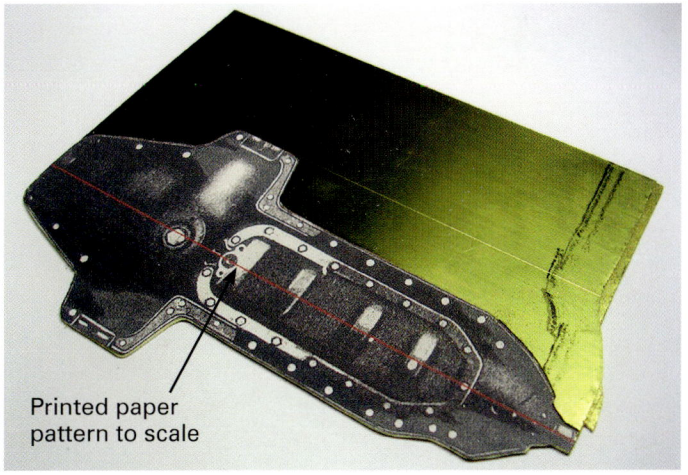

Two sheets of .015" brass spray glued together, then cut out with a jeweler's saw. These will establish the crankcase and oil pan connection.

Here you can see the two pieces separated after having the outer perimeter sawed out using a jeweler's saw and the flange holes drilled around the perimeter.

The inner oil pan opening was cut in the lower piece. The holes were then drilled using the paper pattern as a guide that had been glued back on the lower piece. The paper pattern was saved to use later to make the oil pan itself.

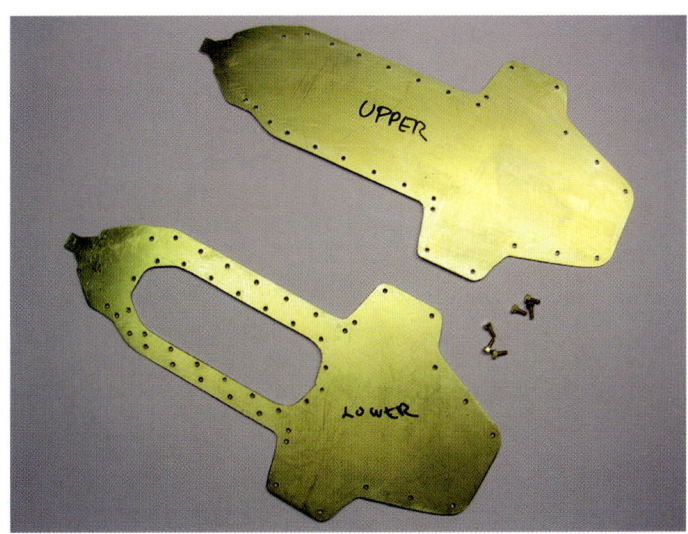

With paper pattern still in place, drill bolt hole locations with #61 drill bit, then cut 00-90 threads. Then cut out the location for the oil pan cover following the paper pattern which is still on the bottom side of the lower piece.

43

Here you can see the perimeter of the oil pan flange has been carefully cut out. The balance of the inner area of the lower piece was cut away leaving just the lower flange that has had the clearance holes drilled out for 00-90 bolts. Trace the saw lines with a sharp scribe to use as guidelines while sawing out the excess material.

TIP: Until you get used to using a jeweler's saw, the brass sheet should be spray glued to a piece of .032" plywood to provide extra strength while sawing to avoid bending or distortion.

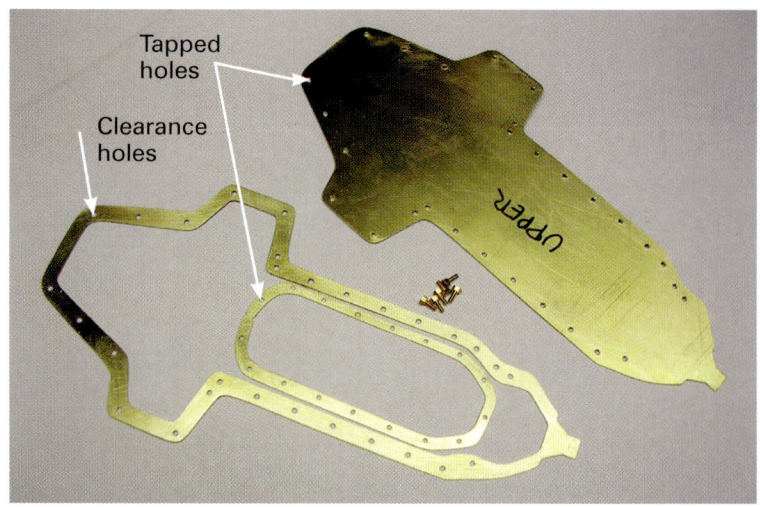

Here the lower oil pan half is drilled, tapped, all excess material cut away ready to build the side.

To build the side walls of the crankcase, I used card or certificate stock available from an office supply store to cut and make patterns for brass pieces that were needed. First, you can start with printer paper to cut and finalize the shape. Once satisfied with the configuration, the pattern is easily traced to the card stock and cut out using scissors. Keep test fitting and adjusting the fit as needed since the card stock replicates the brass sheet. Trace the final card stock shape onto two .015" brass sheets spray glued together, cut out with the jeweler's saw, and file the edges to eliminate any burrs from the saw cutting.

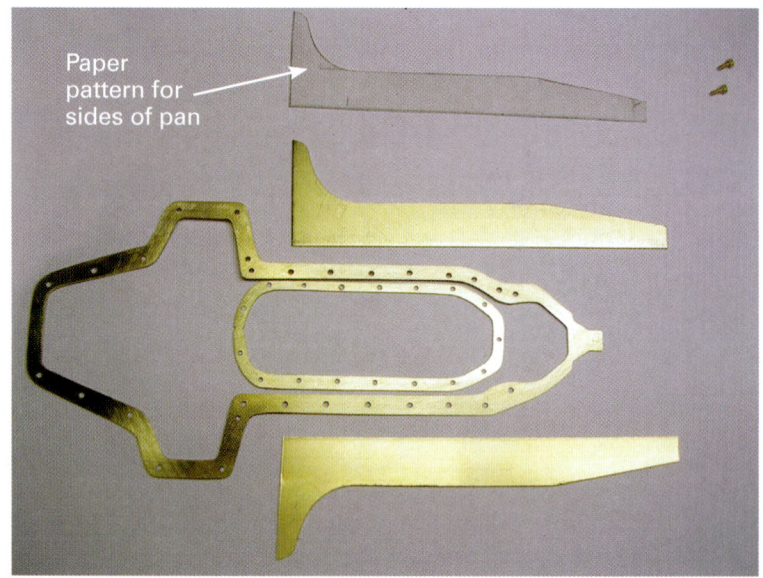

Two identical sides cut from two sheets, spray glued together.

The flange is taped to a smooth, flat surface; I use MDF shelf board with a white laminated finish available from the local DIY store. The brass side is slowly and carefully formed to follow the cut contour of the flange. In this case, I just used my fingers with a .25" rod clamped in a vise. Care must be used to keep a tight fit at the intersection of the flange and side wall. Remember solder likes a tight fit. Note the reference pencil line to keep both sides ending at the same place.

Side pieces were annealed by heating with a torch and then quenching in cold water. Then slowly and carefully form side to fit the contour of the flange piece. Start forming from reference line at rear and move forward, making bends.

Here both sides have been formed and taped in place for the first soldering step at the front where the two sides are butt fitted together for a tight joint.

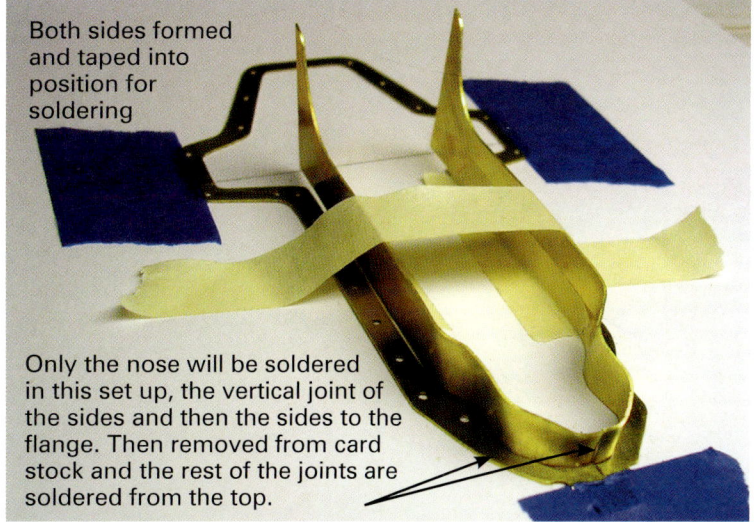

Both sides formed and taped into position for soldering

Only the nose will be soldered in this set up, the vertical joint of the sides and then the sides to the flange. Then removed from card stock and the rest of the joints are soldered from the top.

The solder joint was started at the front with the vertical sidewall joint first, allowed to cool, then down the two sides, making sure that the joints were tight prior to soldering. This joint will assure enough strength to locate both side walls and allow for tape removal.

Nose soldered, not finished up, extra solder was added to form fillet.

The two rear corners were then tack soldered in place, and the joint flowed forward to merge with the front joints on each side. Care must be taken to make sure the joints stay tight and flat. Metal alligator spring clips can hold the flange and sidewalls tight until soldered. This is the critical part that will be the foundation for the rest of the crankcase.

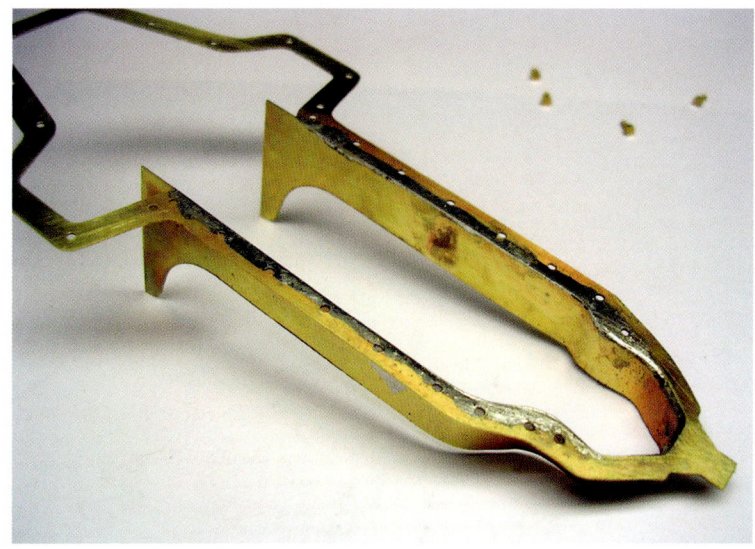

Top view of solder joint, not cleaned yet.

Moving forward a few steps, you can see the oil pan flange with the tapped holes has been soldered in place. This joined the two side walls and provided mating edges for the next parts. Again using card stock, patterns were cut and trimmed for the front and rear surfaces. The rear shape forming the vertical wall was a fairly complex little shape. The brass was annealed by heating and quenching in water, then the top edge was bent using a rod as a mandrel. The lower corners were formed slowly by hand and trimmed until a tight fit, then soldered in place.

Oil pan flange located, front and rear, closed in with hand formed pieces

Rear piece is a compound concave surface transitioning to a 90 degree bend. This kind of piece is why annealing is critical when working with brass.

Here you can see that once the front wall was soldered, a slight crown was formed with small duckbill pliers. Again, using card stock, two identical lateral wall shapes were cut. The side wall was trimmed back using the lateral piece as a guide, then the front wall was cut and trimmed to mate to the side wall. Working slowly and patiently is required. Once both front lateral walls were soldered in place, a "T" wood spacer was added and the rear lateral wall was temporarily tack soldered at the rear lower edges.

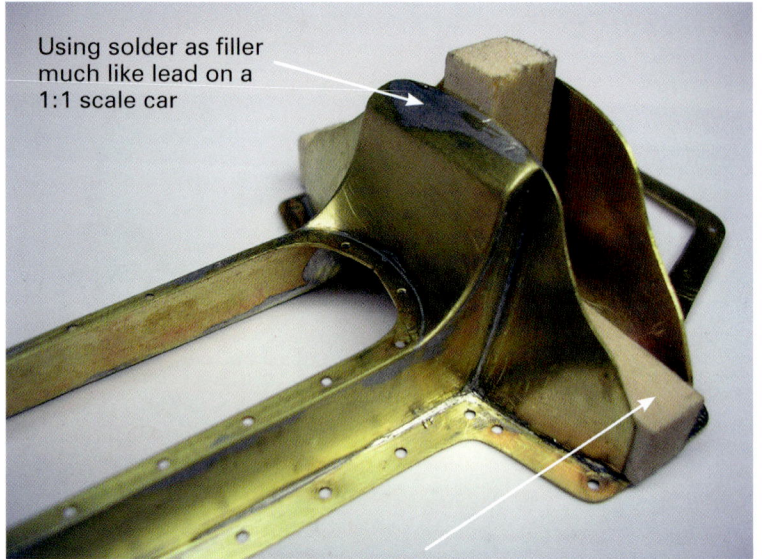

Using solder as filler much like lead on a 1:1 scale car

Wooden 5/16" square T-spacer used to maintain equal distance from front wall while soldering.

The upper flange was bolted to the lower. This keeps the unit flat and provides a surface for the card stock pattern to rest against while fitting and trimming. Prior to bolting the two flanges together, the mating surfaces were both painted with yellow ochre to prevent being accidentally soldered together while soldering the bent shape.

Note the wood spacer is still in place to keep the walls located where needed. This rear cover part will be made and soldered in place to provide stiffness before fitting the side wall parts.

Using CAD (Cardboard Aided Design) to develop pattern for brass. Actually certificate card stock.

Once satisfied with the size and fit of the card stock pattern, the shape was transferred to a piece of brass sheet and slowly and carefully bent to the final configuration. The piece should index into place with slight spring tension to hold it in place and provide a tight fit. The edge against the wall required slight adjustment with files because of the curve in the rear wall.

Brass shape as a result of CAD pattern. Ready to be tack soldered in place.

This shows the finished solder joint before being cleaned up. The soldering was started at the top and worked down each side of the curve. Let the solder cool, and then remove the bolts holding the upper flange on. The unit was placed on a flat surface and taped down. To solder the joint, start at the rear of the curve and move forward to the intersection. Use plenty of flux and a hot iron.

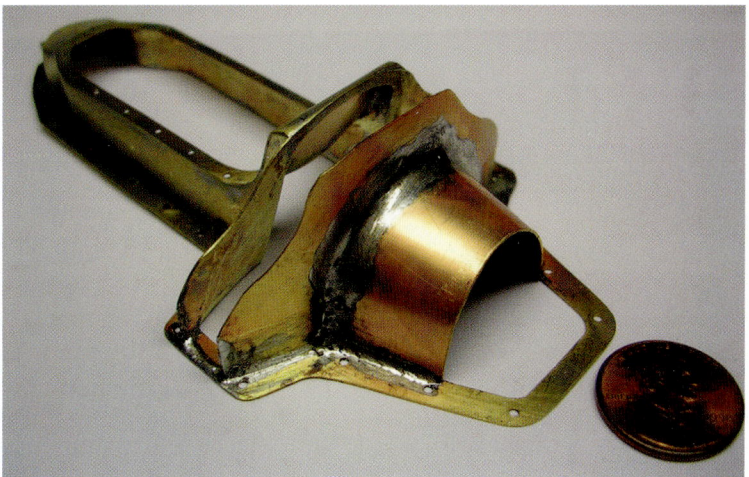

Soldered in place, no clean up yet.

More yellow ochre was applied to the upper flange and bolted back onto the lower flange to hold the unit flat while fitting and soldering the pieces required between the walls. Again, this requires patience to carefully cut and trim while keeping the walls parallel when forming to the contour. These side walls were made with two pieces and the mating butt joint at the top center.

This area not soldered while attached to bottom flange

Soldered while bolted to bottom flange to maintain flatness. Two 5/16" strips were formed starting at the flange and working to the centerline. The rear wall was carefully formed and fitted for a tight joint. The assembly was then removed from the bottom flange and the joint at the flange then soldered off by hand.

The flanges were left bolted together for strength while cleaning off the excess solder and material. Files were used to trim excess brass stock at the sidewalls. The back edge of a #11 X-ACTO® blade was used to scrape the solder away and blend the radii. Once, very close to completion, the part was wet sanded with 320 grit sandpaper, followed by 400 grit sandpaper. All surfaces were buffed with a Scotch-Brite® pad.

Solder joints cleaned up

Now for the tricky part, sawing out the rear sidewall. This was done very carefully with a jeweler's saw. By cutting away several small sections of the sidewall, it was ready for the final trim and blending with a sanding drum on a Dremel® tool. The brass will eat up the drum, so have a few extra sanding drums handy and work slowly to achieve clean mating surfaces. This was easier to do than it sounds. Keep the blade of the saw well lubricated with beeswax to avoid breaking the blade. Have a few extra blades handy as well.

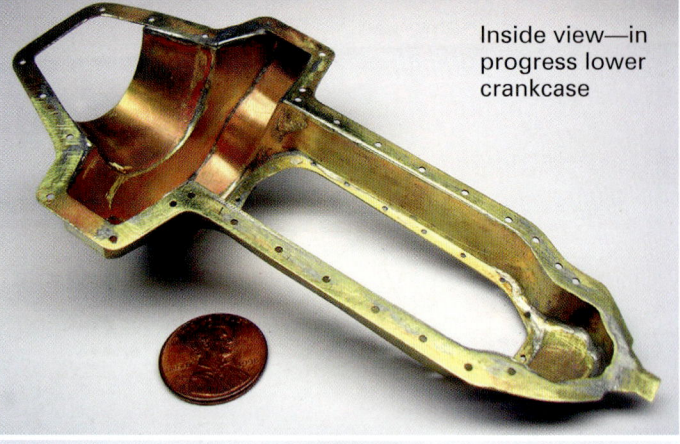

Inside view—in progress lower crankcase

Moving to the rear of the crankcase, three identical parts were needed—two using .020" sheet stock and one using .032"—all spray glued together prior to drilling and cutting. The .032" piece was tapped to eventually become the end flange for the crankcase. The other two were the flanges for the ball socket that enclosed the universal joint.

Here you see the paper pattern used as a drill and cutting guide for all three parts. Pilot holes were drilled for the saw blade to be inserted into to cut out the negative area.

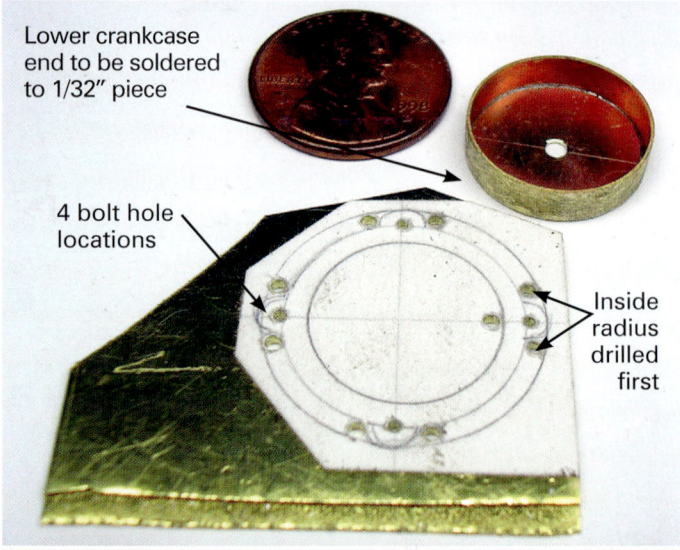

Lower crankcase end to be soldered to 1/32" piece

4 bolt hole locations

Inside radius drilled first

Start of drive shaft ball socket flanges. Three identical pieces needed, one being 1/32" to be soldered to the lower crankcase.

Here, the three flanges are cut out, and the .032" tapped for 00-90 bolts that hold the other two in place. All three were cleaned and dressed with files to create identical shapes. While bolted together, scribe a reference line on all three top center edges for future alignment.

A ring of tube was cut and soldered to a .032" sheet that would become the back wall of the crankcase. The outside diameter was the size of the inner circle with a hole drilled in the center of the diameter of the crankshaft.

Rough cut from sheets, will be final shaped and dressed with files

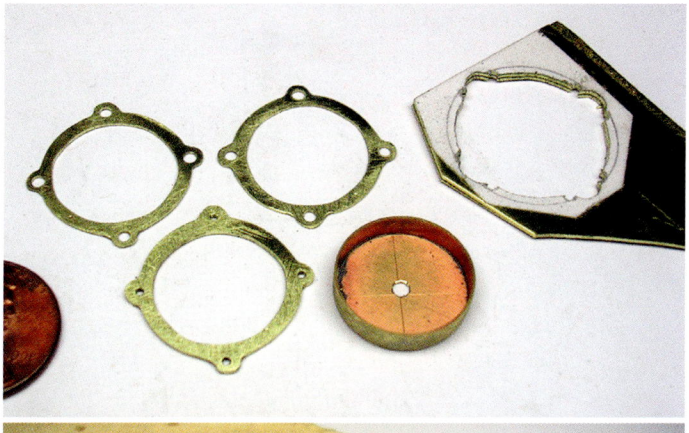

Here is a breakdown of all the finished pieces.

Pieces cut out, lower piece is 1/32" tapped for 00-90 bolts, upper pieces have clearance holes drilled and will become the front and rear of the ball socket.

Cutting in half, once saw cut is made then scribe reference lines for top and bottom flange thickness. Piece stuck to carpet tape to hold while sawing.

The .032" part and ring were then soldered together. The part was then cut through the center using a jeweler's saw. First, a centerline was scribed, then two reference lines above and below the centerline were scribed equaling the thickness of the two crankcase flanges. Once cut apart, each half was filed down to these reference lines and, when soldered to the flanges, would net a perfect circle. Note to keep the top scribed reference line oriented up.

Binding wire to hold half in place while soldered. Middle section of flange will be cut away once top piece is fitted and soldered in place.

After being filed to remove the flange thickness, the top half was aligned to the end of the flange, held tightly in place, and soldered. It was important to get a very good joint on each side.

Once soldered, a crescent moon shape was cut and soldered to the rear edge of the bent shape and a second smaller bent shape piece was cut and soldered to the back of the flange and crescent shape, thus closing the lower crankcase. The flange web across the back was carefully cut away using a jeweler's saw. It was then trimmed and cleaned with a sanding drum on a Dremel® tool.

A triangular rib was cut and bent to form the oil drain shape to the bottom of the crankcase and soldered in place.

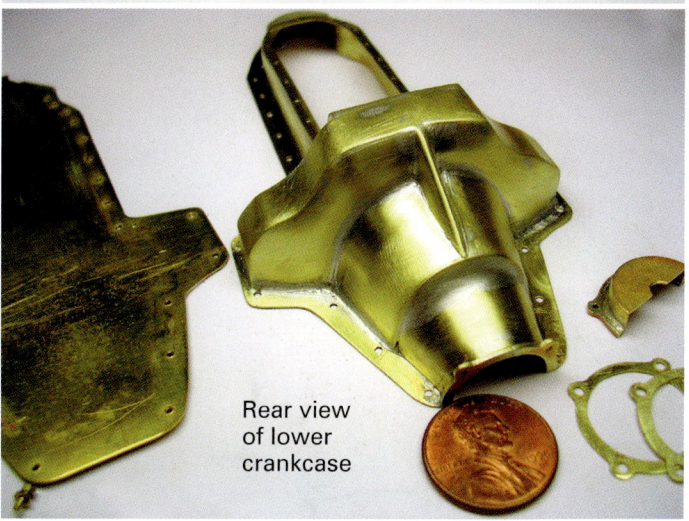

Rear view of lower crankcase

Inside view of crankcase

Here is an inside view of the crankcase to this point. On the left is the flange shape for the upper crankcase, and on the right are the ball socket flanges and the other half of the rear crankcase.

Here is the outside view to this point.

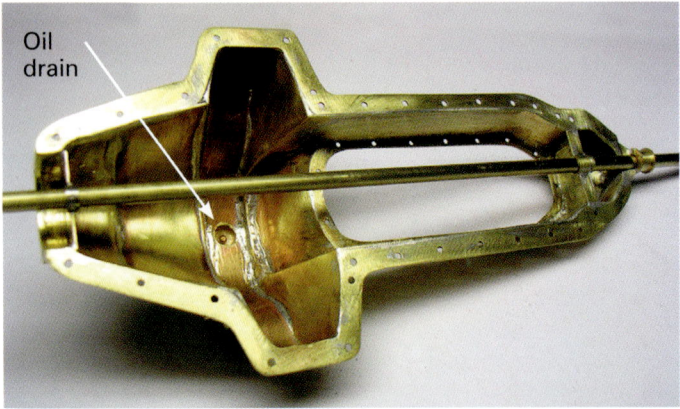

Oil drain

Now, a hole was drilled for the oil drain in the part turned on a lathe. The hole was then drilled and tapped 1-72 thread for the drain plug.

The tube for the crankshaft is fitted with front and rear bearing surfaces added to the standing ribs.

Front engine wall added with alignment support added. The upper half will be sacrificed once the upper engine block is built, for now it is used as a guide.

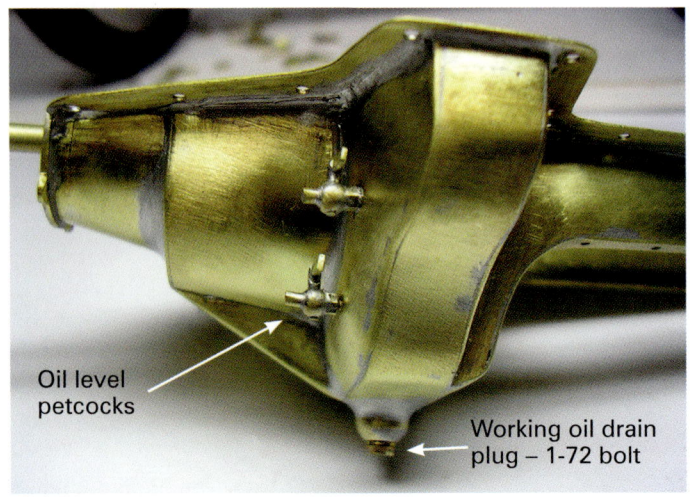

Oil level petcocks

Working oil drain plug – 1-72 bolt

The oil level petcocks were fabricated and added.

All excess and undesirable solder was removed using needle files, sandpaper, and Scotch-Brite® pads. For radii, use the butt end of a drill bit the proper radii size as a scraper. The last step was to polish with a toothbrush, pumice powder, and water to get an overall smooth finish. This step may reveal more areas that may need additional scraping, filing, and sanding. The cleaner and smoother the finish, the better the end result will be, especially if it is to be electroplated.

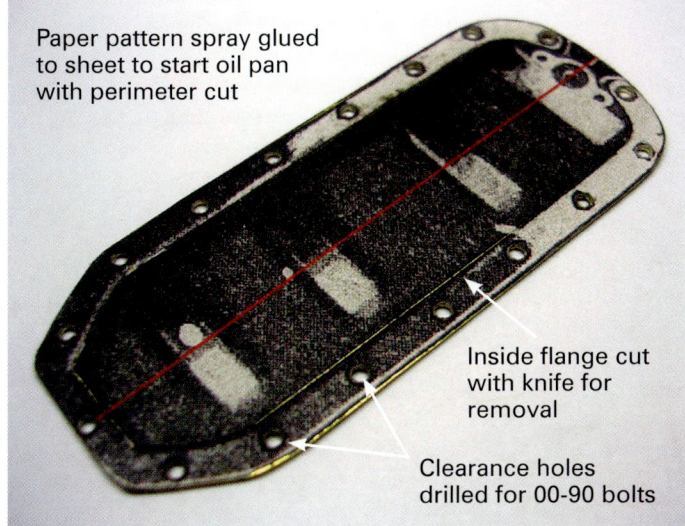

Paper pattern spray glued to sheet to start oil pan with perimeter cut

Inside flange cut with knife for removal

Clearance holes drilled for 00-90 bolts

Now moving on to the oil pan cover. Here is the cut out part with the paper pattern still in place. With a sharp X-ACTO® blade, the inside perimeter of the flange was cut and drilled for the clearance holes for the 00-90 bolts. Remember to center punch the starting locations for the drill bit. A small center punch can be made by taking an eyeglass screwdriver and grinding the end to a point. This center punch can then be twisted at the location points to be drilled. Because brass is relatively soft, this tool works great. Then, carefully with a Q-tip soaked in lacquer thinner, wet the center of the paper pattern to release the spray glue. Take care not to remove the outer flange pattern.

Center area removed so as to use edge as a guide for sawing with jeweler's saw

Pilot holes drilled to insert saw blade

With the center of the paper pattern removed, two pilot holes were drilled to insert the jeweler's saw blade. Using the bench pin as a work surface, the inner shape was carefully sawed out using the paper pattern as a guide and plenty of beeswax as saw blade lubricant.

Remember to save the cut out center paper pattern for use later on to locate the oil well shapes.

Wood forming buck made from basswood sheet. Center area cut out on top piece using flange as a pattern. Top sheet Super Glued to bottom sheet, then flange spray glued to top sheet for hand forming edge. Brass sheet was annealed using torch and carpet taped in place.

A wood forming buck was made using .187" thick wood sheet with the center area cut out, then glued with Super Glue® to a second sheet. The brass flange was then Super Glued to the top piece of wood. By keeping the flange on the wood piece, you will maintain a sharper edge during the forming process. A piece of brass sheet was cut to size and cleaned with a Scotch-Brite® pad. Using the Scotch-Brite® pad helps to see the surface changes during forming. The needed piece of brass sheet was annealed to soften for forming.

When ready to form, double-sided carpet tape was applied over the entire wood surface, and the excess tape was cut away around the inside perimeter of the brass flange. Center the annealed sheet on the forming buck and press the sheet firmly down onto the carpet tape. Keep in mind that enough gripping surface needs to be around the flange to hold the sheet in place during the forming process. Starting in the center of the piece to be formed, press down with a round wooden shape to depress the brass and move slowly toward the outer edges. Use a .25" round hardwood dowel with one end cut at a 45-degree angle to firmly form the tighter perimeter edge.

Brass sheet carpet taped in place and press formed very slowly using a wood handle (round end) and an oak rod cut at a 45 degree angle.

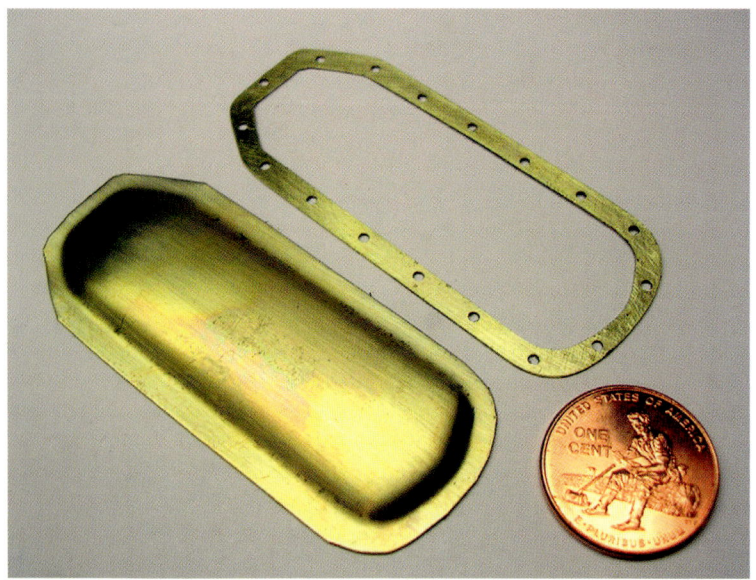

Rough pieces removed from wood buck using lacquer thinner to breakdown the carpet tape adhesive.

The previous forming step was continued until satisfied with the size of the radius. By rotating the dowel during the forming, you will have various arcs on the end of the dowel to work with to get the needed radii. Make sure to also work the upper flat surface to maintain a sharp edge.

Here the finished piece has been removed from the carpet tape using lacquer thinner, and the outer perimeter trimmed to fit the flange. Notice that because of the forming, a slight radius is on the formed part. This is normal for hand forming.

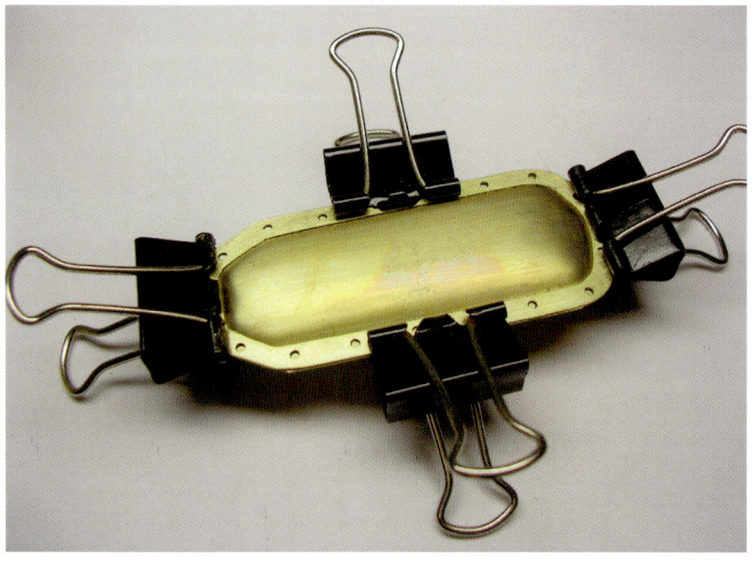

Parts clamped together ready for soldering.

A flange was soldered to the formed oil pan. Apply flux before clamping. Metal spring clamps held the parts tight. With a very hot iron, solder was applied to the edges between the clamps. Watch for the solder to appear at the inner edge of the flange. See that enough solder is in place for a strong joint. Move the clamps to the soldered areas, and finish soldering the flange completely. Solder should fill to the edge around the formed pan. Excess solder was cleaned off the flange. The clearance holes were then drilled through.

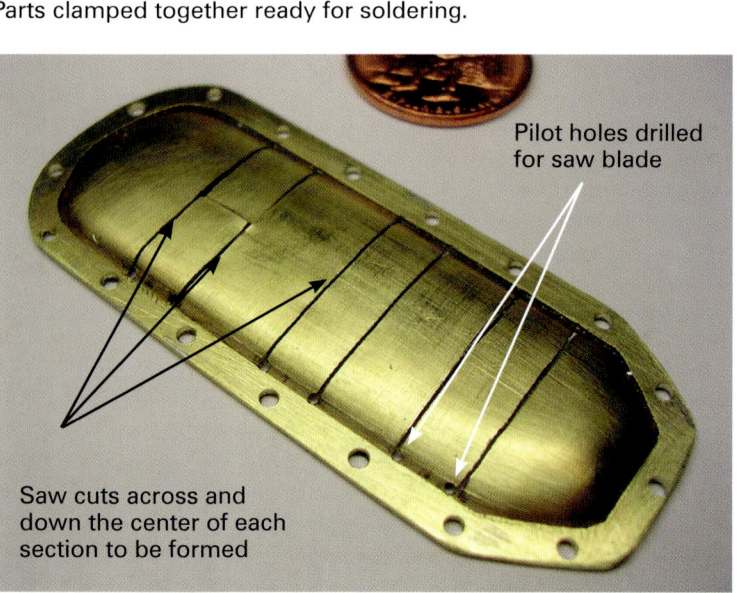

Using the paper pattern for the center, locate and scribe the lines for the oil well shapes. At the intersection of the oil pan and flange, pilot holes were drilled for the jeweler's saw blade to be inserted. Using the guidelines carefully saw across the pan. Once all the guidelines were sawed across, then saw down the center line of each section.

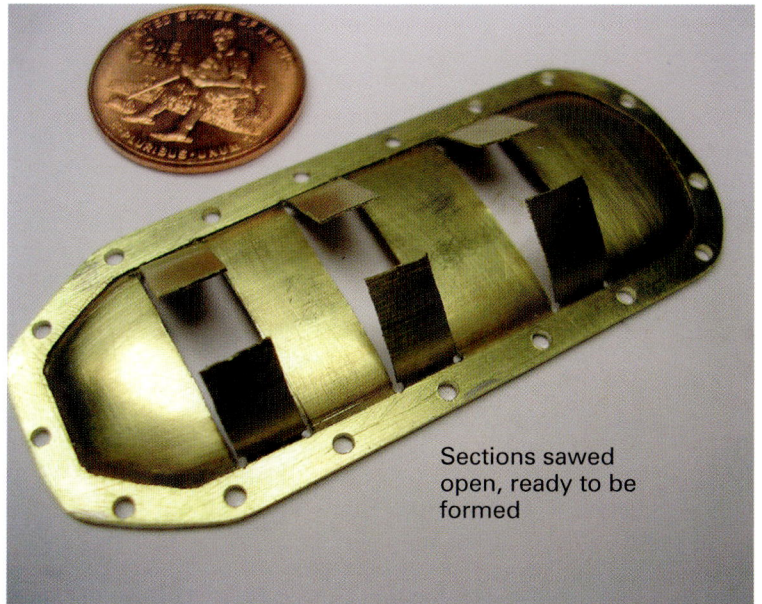

Sections sawed open, ready to be formed

Carefully bend each section up equally using small duck bill pliers to form the curve needed on each section. Cut and form a cap piece to be soldered to the outside of each bent section. This part provided the smooth outer surface and covered the pilot holes at the flange once soldered in place. While cutting the three cap pieces, remember to allow for the draft angle, meaning that each edge will have a slight curve to the center when soldered in place. See associated picture.

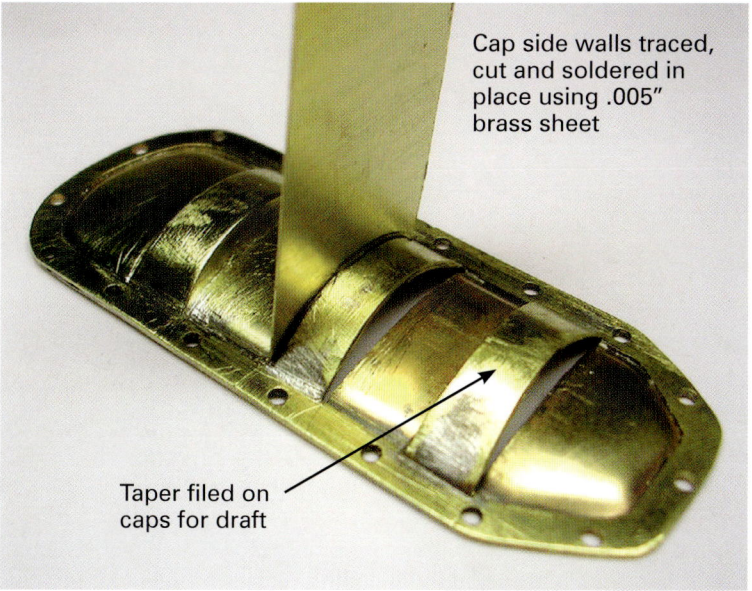

Cap side walls traced, cut and soldered in place using .005" brass sheet

Taper filed on caps for draft

Once each well was cut and filed to allow for draft (angle upward), a .005" thick sheet stock cut to width was placed against the well edge and traced with a scribe, then cut and soldered in place. Again, the tighter the fit, the less filing and cleaning is required.

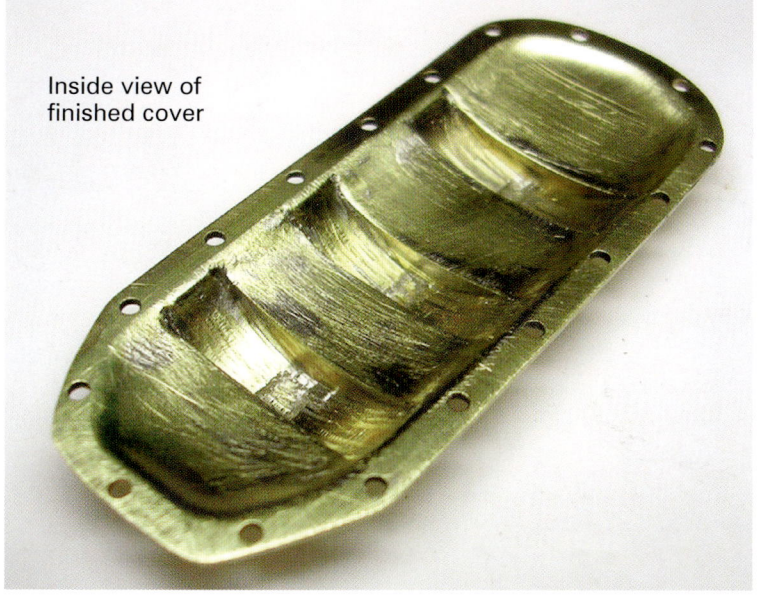

Inside view of finished cover

Here is an inside view of the finished oil pan cleaned of excess solder and ready to install on the lower crankcase.

The fabrication process used to build this oil pan can be used to form various configurations of parts and components. The lesson here is the use of multiple layers of brass sheet stock to attain the final desired shape and is a quicker building alternative to machining.

The finished oil pan is bolted in place to the lower crankcase half. My hand gives it a sense of scale, but, from front to rear, the assembly is about 5" long and 2 3/8" wide.

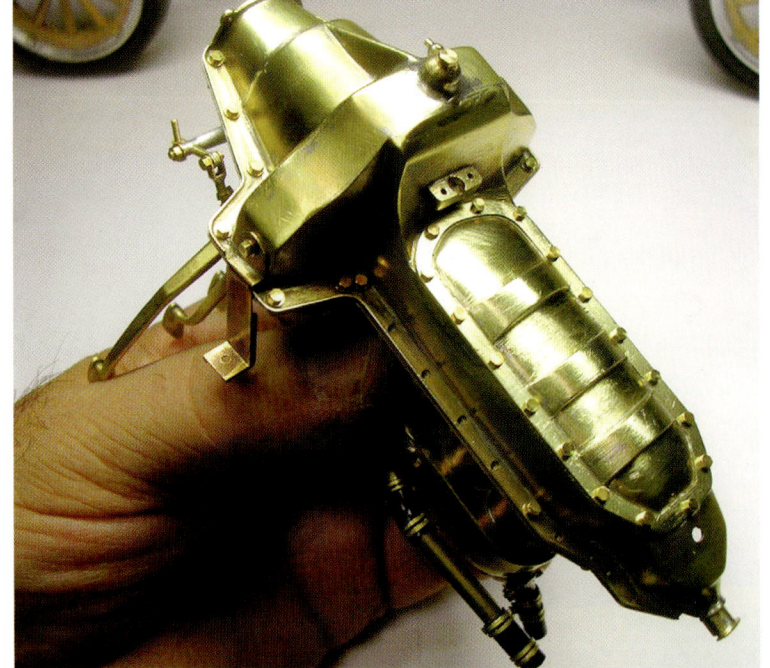

These are all the basic tools needed to clean excess solder off parts after soldering. Of course, the initial goal is to use as little solder as necessary and still have a very good strong joint. When the part is cleaned to satisfaction, the next step would be to polish the part. To polish, use a hard bristle tooth brush with powdered pumice and water. Scrub over a bowl of water set in a sink. The bowl collects the excess pumice during rinsing to avoid plugging the sink drain. This polishing will also reveal areas that may need additional clean up.

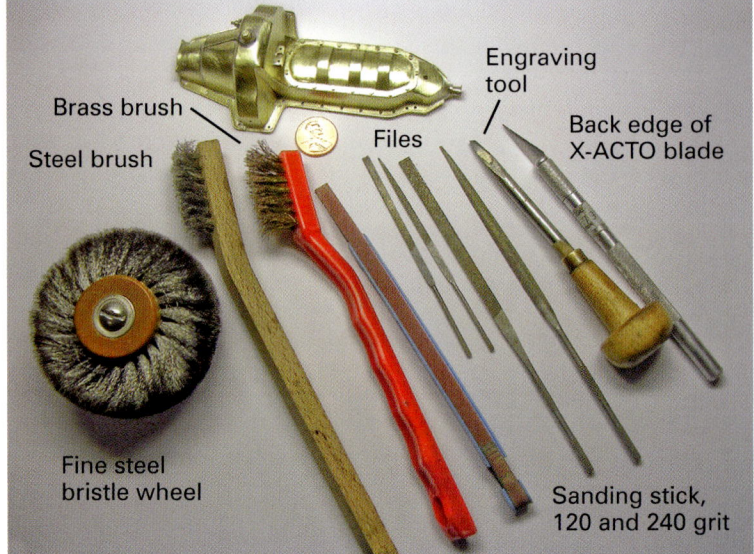

Part cleaning tools.

The lower crankcase is cleaned, polished, and ready for electroplating.

This is about as complicated an assembly as may need to be formed. With careful planning and hand tools, this crankcase was completed with a nominal wall stock of .015" sheet stock for the needed clearance for the crankshaft to turn the pistons.

To this point, all shapes were fabricated by hand using hand tools. The holes were drilled using a tabletop drill press. The upper half of the engine requires the use of machine tools.

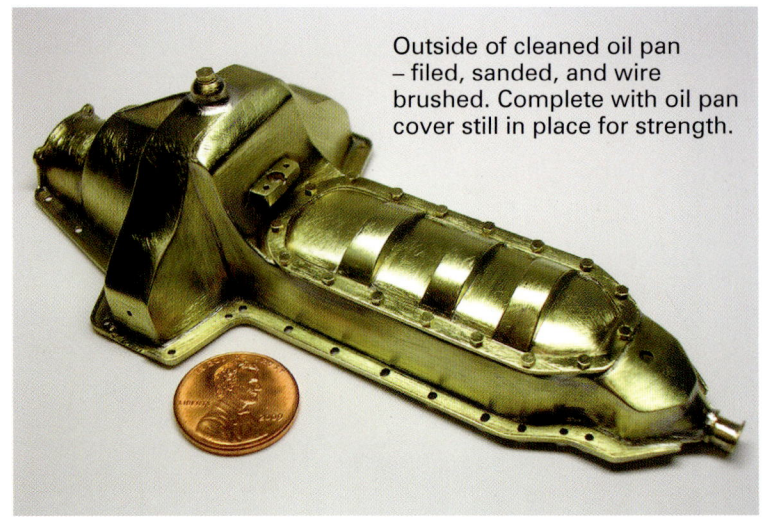

Outside of cleaned oil pan – filed, sanded, and wire brushed. Complete with oil pan cover still in place for strength.

Part Fabrication: Incorporating Round Tube

Now, to move on to the upper portion of the engine. This fabrication will get more involved and require the use of a lathe to cut multiple tube parts to identical lengths. While tubes can be cut by hand, these cutting operations are easier to accomplish with a lathe using a cut off tool.

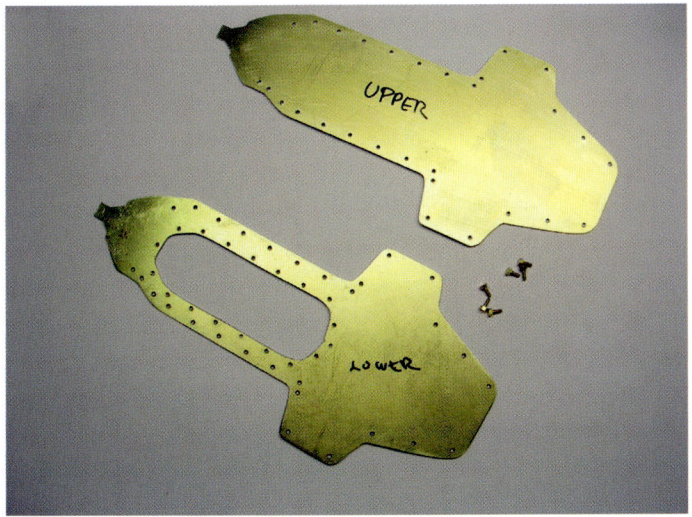

With the paper pattern still in place, drill bolt hole locations with a #61 drill bit, then cut 00-90 threads. Then cut out the location for the oil pan cover following the paper pattern, which is still on the bottom side of the lower piece.

Using the finished lower crankcase, identify which side of the part is up and mark it for ongoing reference. This is important when working with asymmetrical parts. The paper pattern for the cylinders was cut and spray glued to the up side of the cut out flange part. Make sure to line up the holes in the proper location.

Note the web section to hold the center crankshaft bearing, which was cut around, leaving it in place. Pilot holes were drilled for reference later on for the bearing brackets. The pilot holes were drilled to insert the jeweler's saw blade. Once both sections were sawed out, they were cleaned and dressed with a file to true up to the final shape.

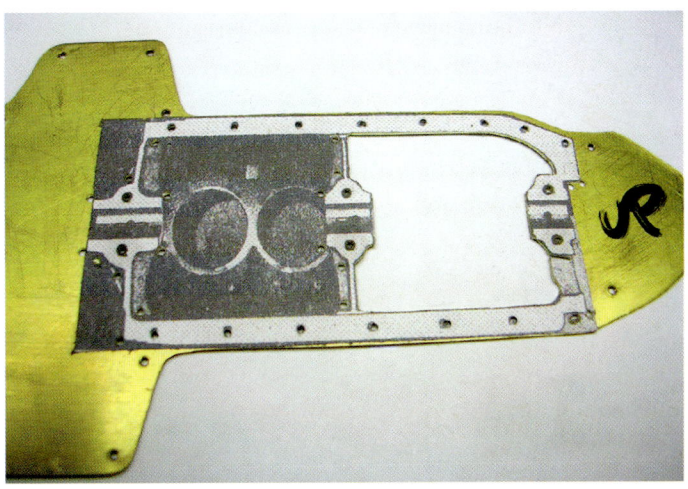

Here, you can see both sections cut out and edges trued up with needle files.

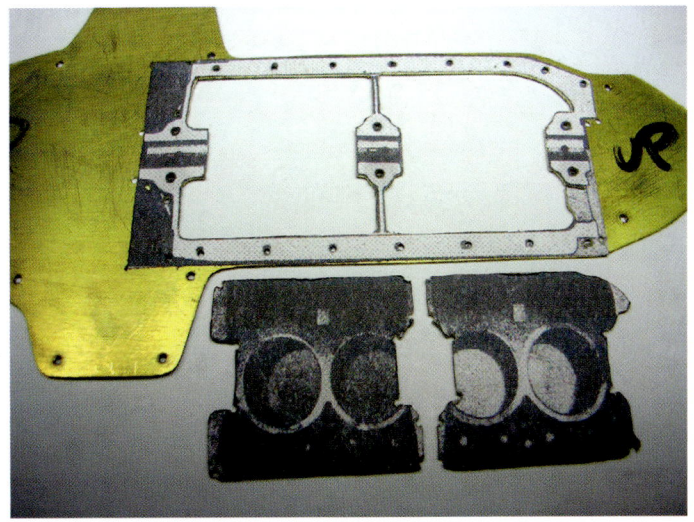

This shows the finished flange with the bearing mounts still in place for future reference. The flange is bolted to the lower crankcase and the perimeter is trued up using files, so that both parts match each other.

This flange part will be the foundation for the upper half of the engine. Prior to bolting, apply yellow ochre to both mating surfaces and bolt threads to prevent accidental soldering together of the parts during later soldering operations.

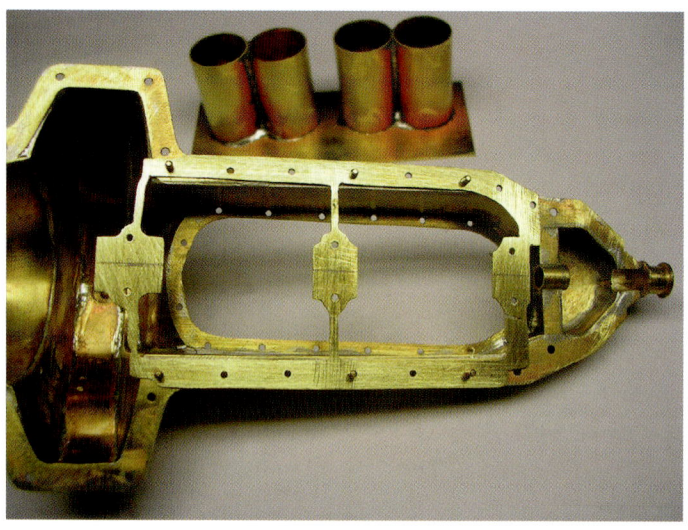

A brass sheet was scribed with the length and width of the lower flange part with the length being the distance from the outer edges of the bearing mounts.

Holes were drilled in the location for the piston cylinders to press fit into.

TIP: To drill holes in small parts like this, it is best to adhere the part to a piece of hardwood with carpet tape to hold securely while drilling.

Two cylinders were located, bound with binding wire, soldered to each other and to the brass sheet.

Three sheets of brass were cut using the layering method to have three identical parts. Subsequently, the parts were drilled with required holes while still spray glued together. The large holes are for the cylinders, the medium size holes for the push rods and the small holes for the head bolts. The lower part in the picture was the test part and used later as a pattern to trace the copper head gasket.

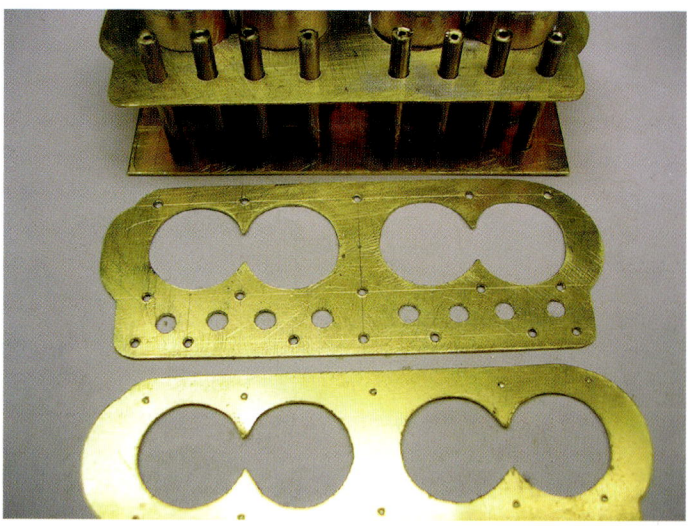

Here is the breakdown of the parts needed for the upper engine assembly. The push rod hole locations have been added to the bottom layer sheet part.

The four identical taller tubes in the middle were cut on a table top lathe and have been sawed vertically to slip over the cylinder walls. They will function as indexing shoulders for the second layer part. The shorter tubes will be spacers for the top layer part.

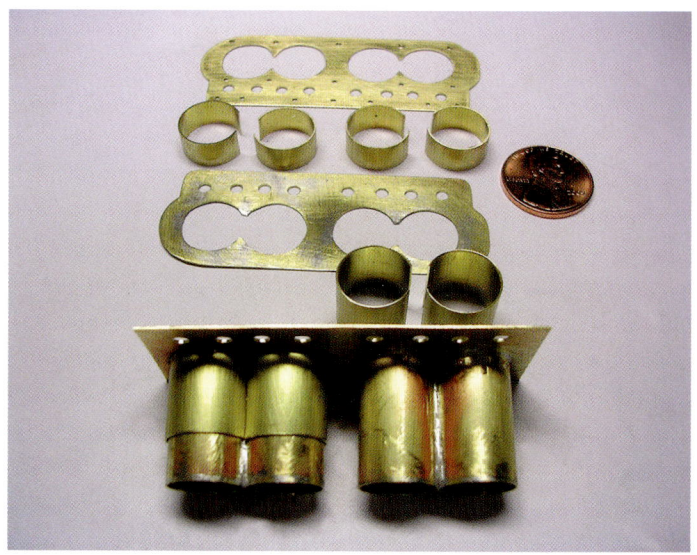

Here you can see the taller tubes slip fitted over the cylinder walls providing a good resting shoulder. Trimming and filing was required to get a good tight fit of the parts to one another.

In the foreground are the pushrod tubes cut to length from top to bottom.

Here the pushrod guide tubes were cut and test fitted. This shows the value of making identical parts using the layering process. Reamers were used to remove fractions of metal to accomplish a press fit for the tubes into the three layers of holes.

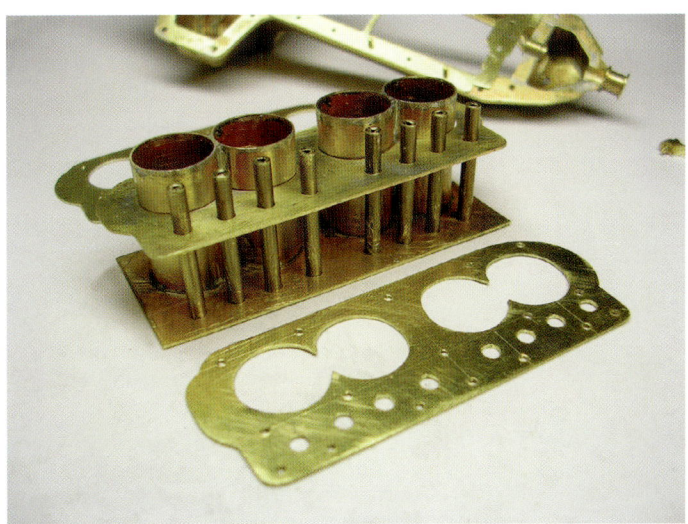

This shows the second layer sheet brass part press fit in place onto the taller tubes. The shorter tubes were slip fitted over the cylinders in the same manner as the taller ones.

This picture shows all three layer parts in place for soldering, and, once clamped, all the parts will stay in place because of the shouldered tubes. The wood blocks acted as spacers to allow room for the 120 watts soldering iron. Again, plenty of soldering flux and a very hot iron were used. A propane torch could be used if you feel comfortable using one. Be careful to rotate the assembly while soldering to get good solder flow and joints. The clamp was a holding handle while rotating to allow the liquid solder to flow where needed. Remember solder flows to heat and reacts to gravity.

After the soldering operation was completed, it was allowed to cool, then cleaned of excess solder before proceeding to the next step.

The pushrod guide tube spacers were all cut to length, three for each pushrod guide tube. The pushrod guide tubes were then inserted into the bottom holes and, as advanced, two of the short tubes were slip fitted onto the tube, then inserted through the second layer. The longer spacer tube was slipped on and the tube press fitted through the top hole. Once all were in place, they were then soldered to the layers and each other. This assured perfect alignment for the pushrods and valves.

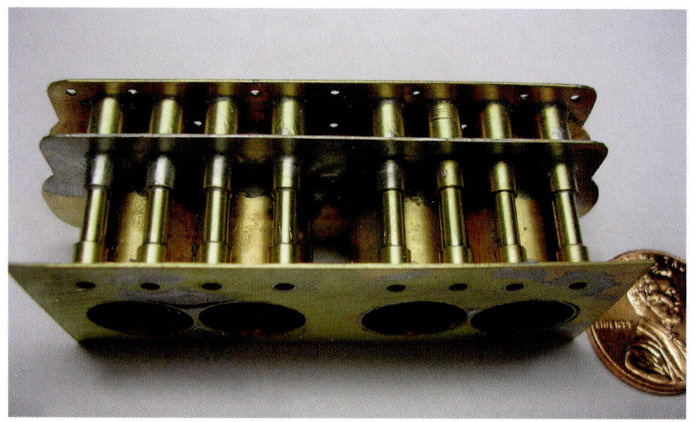

Excess solder was removed. Then, using the tube shoulder as a guide, the tubes were cut between the upper and lower guides. The saw cuts were filed flat and true to each other. The holes were reamed to remove any burrs on the inside.

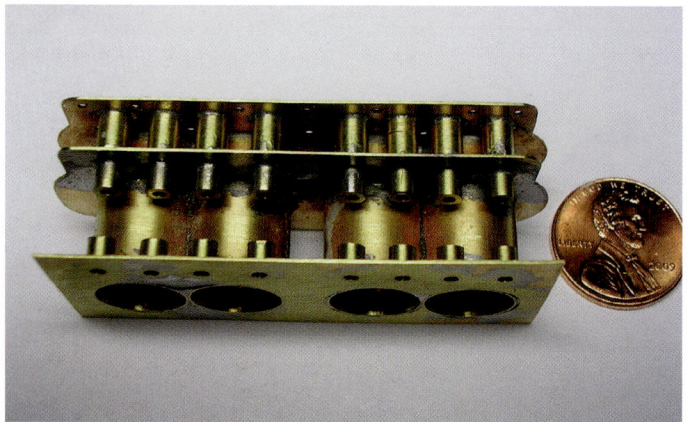

Here the water jacket sidewall has been formed and drilled for the intake and exhaust locations between the two upper layers. This sidewall was added all the way around to form the upper water jacket. This is the right-side view of the engine block. Here you see the valve seats that have been countersunk with a tabletop drill press.

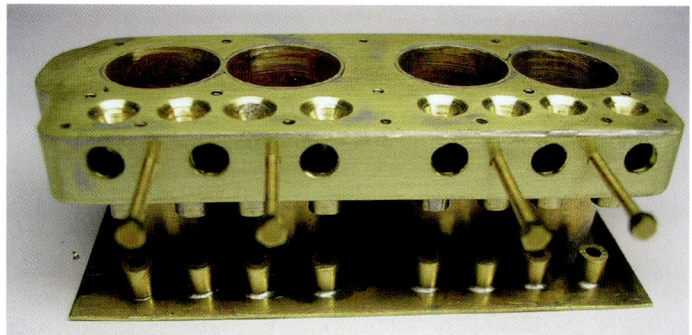

This is the left side view with the hole location for the water return outlet to the radiator. Note that the sidewall is formed using one long piece with the butt joint at the rear end of the engine. This is an example of thinking ahead to how and where seams should be located to be less obvious on the completed parts.

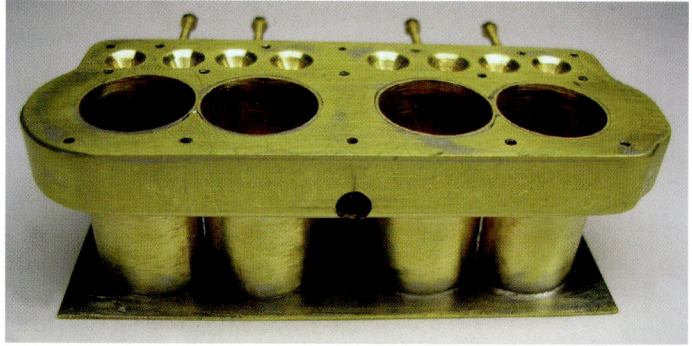

In this picture, you can see the two "U" formed pieces with access holes cut out of the center sections and soldered in place to the cylinders. The two bolt hole locations between the cylinders were drilled and tapped for the cover plate 00-90 retaining studs.

Next, the upper assembly is located on the flange bolted to the lower crankcase using wood blocks to determine the correct height to the top surface. The assembly is starting to look like an engine.

The inside joints were cleaned. The assembly was located on the flange in the proper position. This is where the yellow ochre painted on the flange surfaces came into play. The pieces were clamped together with the sidewalls tack soldered to the flange. The bolts were removed, and the flange joint was soldered to finish the joint on the inside.

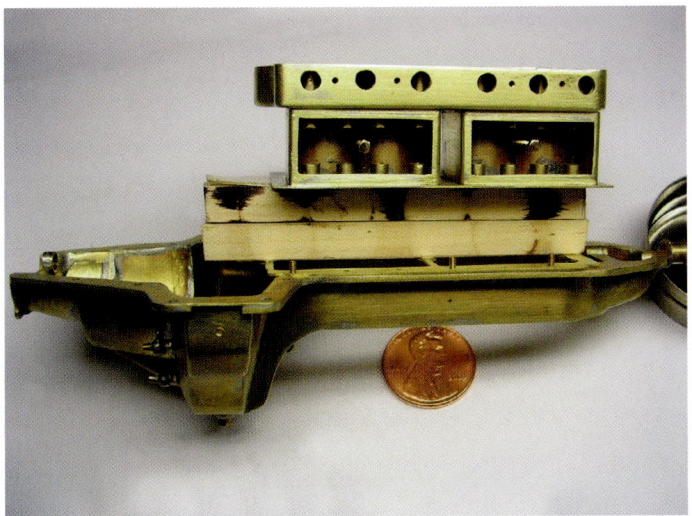

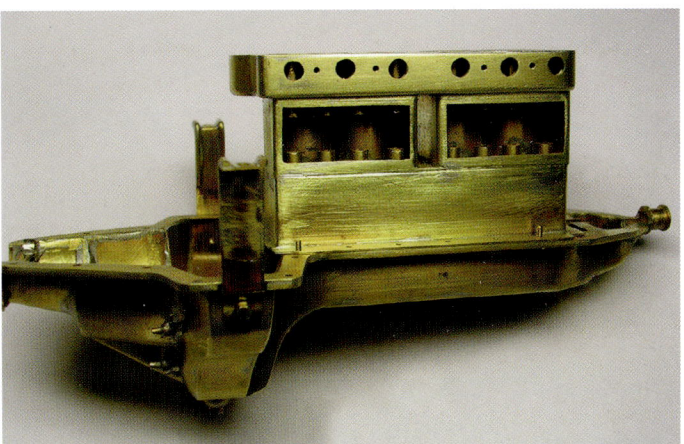

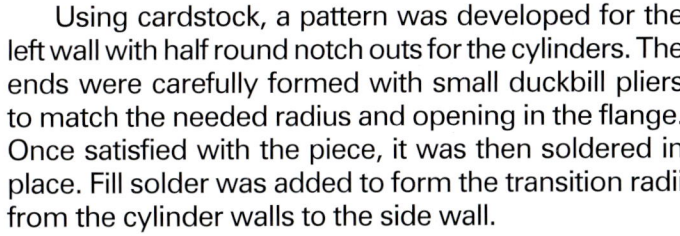

A cardstock pattern was developed for the lower sidewalls with the left side curve cut on each end. This part was formed with corner bends providing the required radii. Then, using one end as a template, the center piece was traced with a scribe, cut, and soldered in place.

Using cardstock, a pattern was developed for the left wall with half round notch outs for the cylinders. The ends were carefully formed with small duckbill pliers to match the needed radius and opening in the flange. Once satisfied with the piece, it was then soldered in place. Fill solder was added to form the transition radii from the cylinder walls to the side wall.

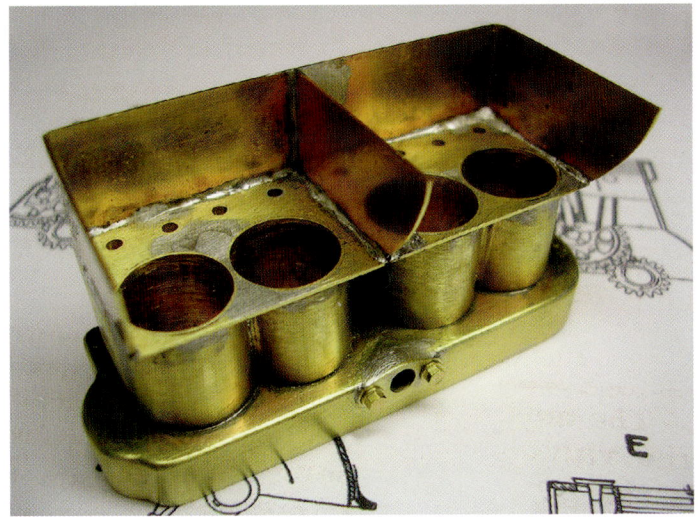

The finished assembly has been cleaned and polished with a Scotch-Brite® pad.

This picture shows a close-up view of the front left formed corner radius. Note the tight butt joint on the curve; this is the kind of fit that is needed for a strong joint in this situation.

Here you can see the bolt pads have been added at each hole location, drilled, and threaded for 00-90 bolts. They also reinforce the sidewall to flange joint. The pads were cut to length with the ends rounded with a file. The hole locations were marked and drilled using a #61 drill bit; the same as in the flange. Each pad was indexed to the flange using a toothpick, then soldered in place.

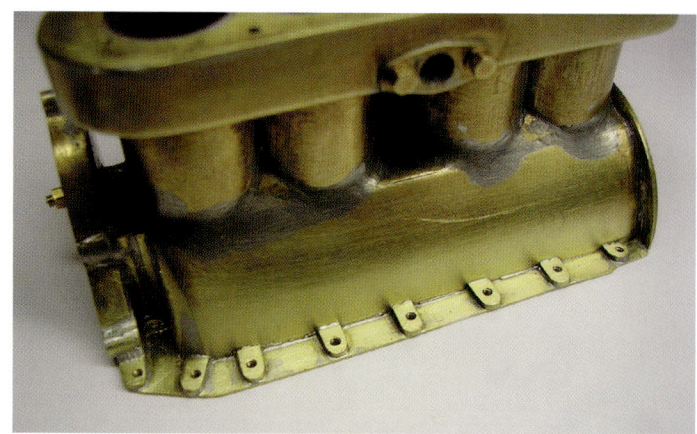

The following sequence will show the fabrication of the transmission cover. Again, this needed to be only wall stock thick since the transmission would turn within it. Start with a card stock pattern. This transmission cover part mated up against the rear of the engine assembly and bolted on. It will be a painted part since the original was cast aluminum.

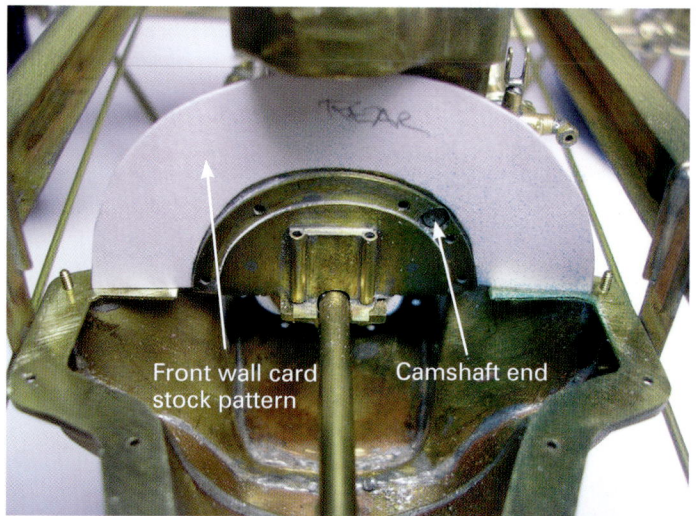

Prior to soldering all mating surfaces of lower flange were painted with yellow ochre to prevent any accidental soldering.

This is a picture showing the lower crankcase flange painted with yellow ochre to prevent the two flanges from being soldered together by accident. Think of yellow ochre as painting on "dirt" so the solder will not stick.

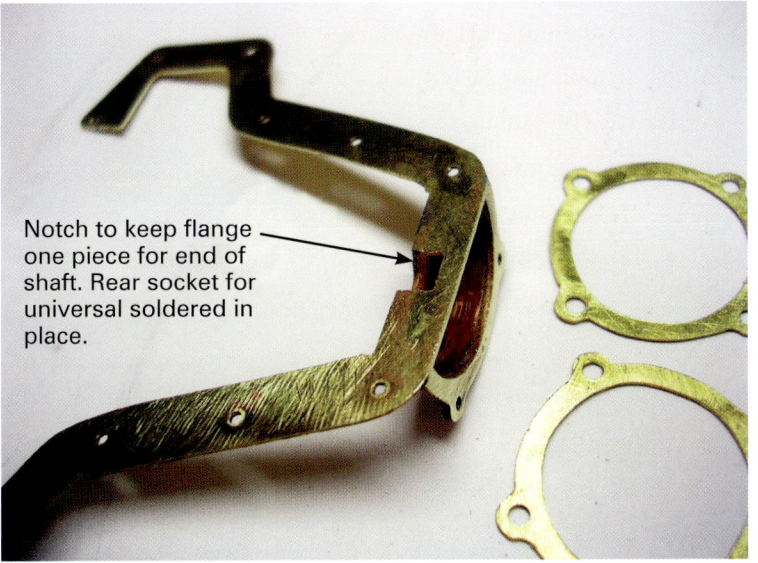

Notch to keep flange one piece for end of shaft. Rear socket for universal soldered in place.

The rear universal flange was soldered on while the cover flange was taped to a flat surface, then the yellow ochre was painted on the flange. Yellow ochre can be purchased from an art or jewelry supply outlet. Yellow ochre comes as a powder and is mixed with a little water to a paste consistency.

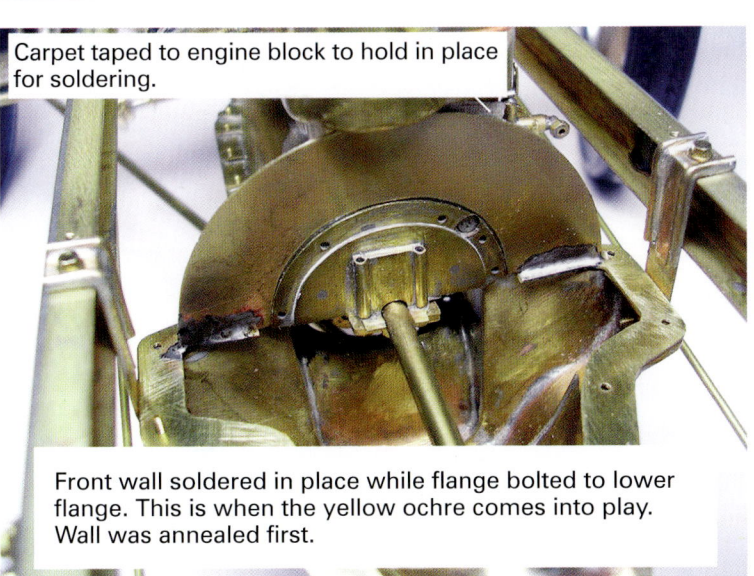

Carpet taped to engine block to hold in place for soldering.

Front wall soldered in place while flange bolted to lower flange. This is when the yellow ochre comes into play. Wall was annealed first.

The cover flange was bolted in place, and the front cover brass piece was soldered to the flange. The front cover piece was annealed prior to being soldered in place. This soldering operation demonstrated the need for yellow ochre so as not to solder the parts to the lower crankcase while bolted to it. It is critical for this to be a good, strong joint because the upper rim of the part had a radius formed while still bolted in place.

While still bolted in place, the rear plate was cut and soldered to the rear universal flange.

Starting at the flange and slowly working to the top center, the radius was formed with small duckbill pliers. Leaving the flange bolted in place helped with strength, while the radius was being formed.

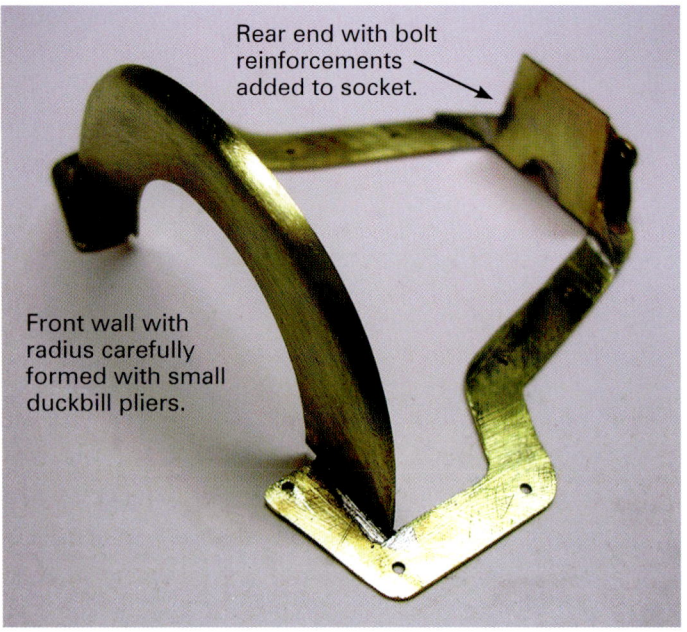

Rear end with bolt reinforcements added to socket.

Front wall with radius carefully formed with small duckbill pliers.

Here, the second rearward piece was cut and formed, however, on this piece, the radius was formed off hand (not bolted in place). Note that less material was cut out of the center area than on the front piece to allow for the angle and lower radius at the flange. This was a tricky, little part to form but a good example of thinking ahead. This piece was annealed for easier forming of the larger compound curves. Note that the flange was carpet taped to a flat piece of shelf board during this phase of the building process.

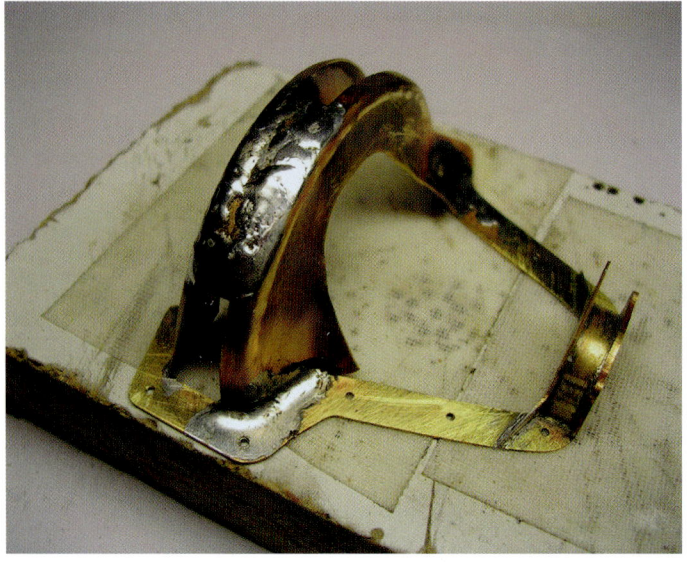

This shows the top part with the excess solder cleaned off. The lower part is being fitted. This part has a slight flare at the bottom as a result of the forming of the radii. The top arced pieces were of equal width and thus kept the side walls parallel to each other. It is critical that butt jointed filler parts like this be a tight fit and pressed into place for the strongest and smoothest blend of surfaces.

The U.S. penny gives a sense of scale.

Cleaned up upper panels.

Right side lower panel cut, formed, and press fit into place.

Here you can see the value of tight fitting butt joints. Once cleaned of excess solder, sanded, and polished, you can barely perceive the joints.

Here is the inside view cleaned up with the butt joints clearly visible inside the part.

Here you can see the top plate is being test fitted to assure correct geometry. Note, the formed side wall top edge trimmed to support the top plate.

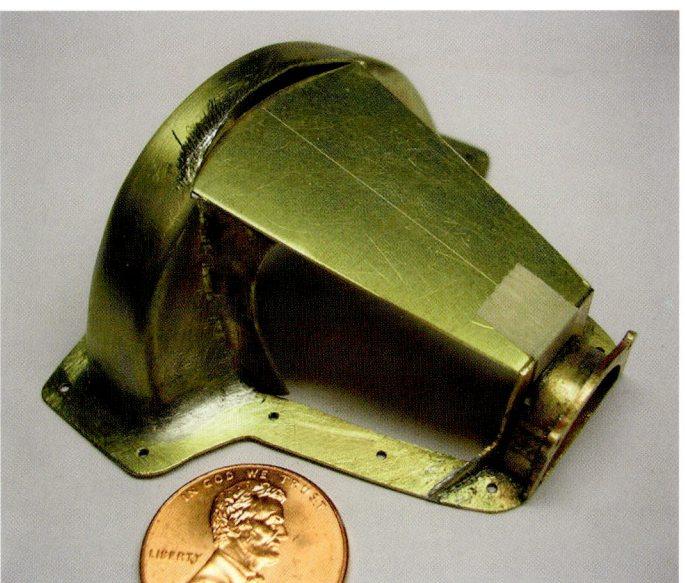

In this picture, after completing the shape, it was necessary to saw out a notch at the top, then cut vertically on the centerline. With the small duckbill pliers, the radius was formed from the flange to the top corner of the notch starting the sidewall that will be trimmed to provide an edge for the top plate piece to be soldered to.

The left sidewall was cut and soldered in place. The top plate can be seen with the access hole cut in it along with the holes drilled and tapped for 000-120 cap screws.

Also note, the small crescent shape has been soldered in place at the top of the transmission cover assembly.

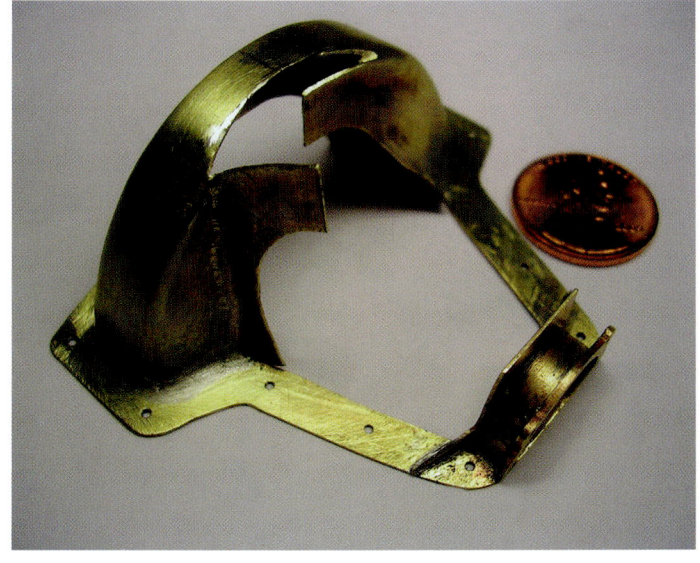

This shows the right sidewall being traced onto cardstock to make the pattern for the right-side brass part. You can also better see the crescent shape at the top of the cover better in this picture.

Here is the finished assembly with the top access cover bolted in place with the 000-120 cap screws.

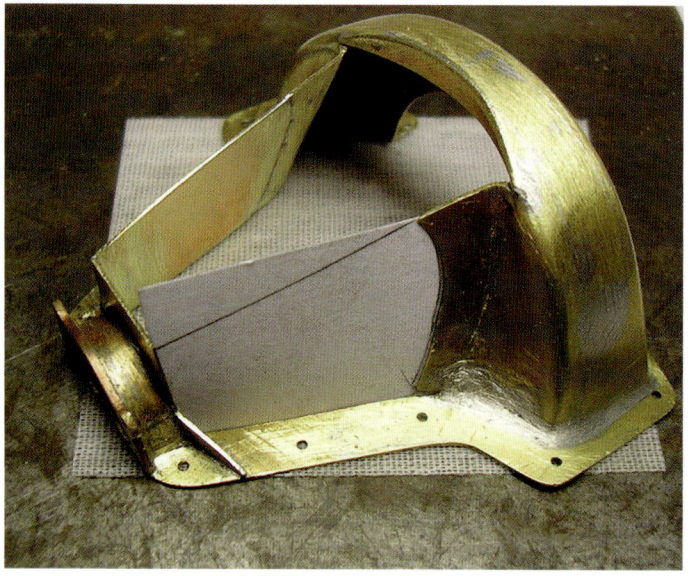

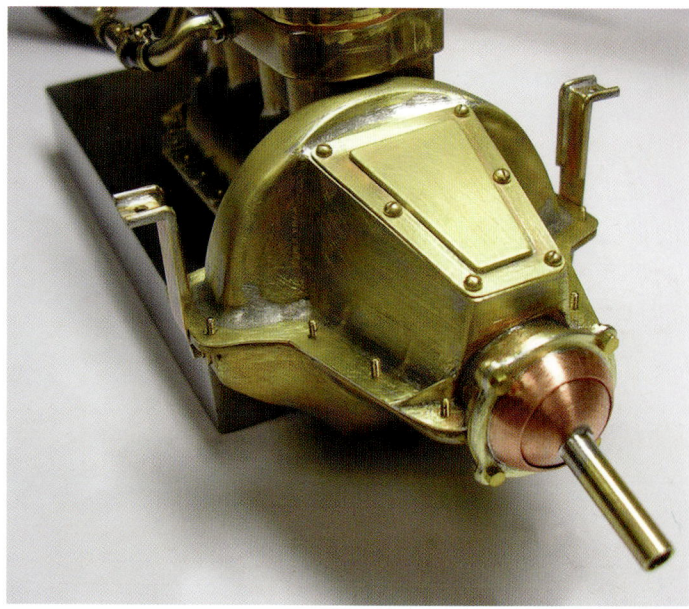

This picture shows the final transmission cover assembly enclosed, ready to clean off all excess solder. As standard practice, clean and polish all components once a certain milestone like this is reached. It is easier to access all joints and neutralize any residue flux that may remain on the part. If left, it would continue to etch the finished surface. Acetone does an effective job of cleaning brass parts without leaving a residue on the surface like lacquer thinner does.

Once the assembly was cleaned to satisfaction, the pedal support tubes were located. Here, the hole locations are located on both sides and drilled to the needed sizes. Lengths of tube were inserted into the holes across the unit and spaced as needed. The holes were filed slightly elliptically because of the angle of the side walls. A press fit was the goal. The tubes were marked on each side of center for the gap needed for the transmission bands. Once marked, each tube was removed and the marks cut on a lathe almost completely through the tube wall stock.

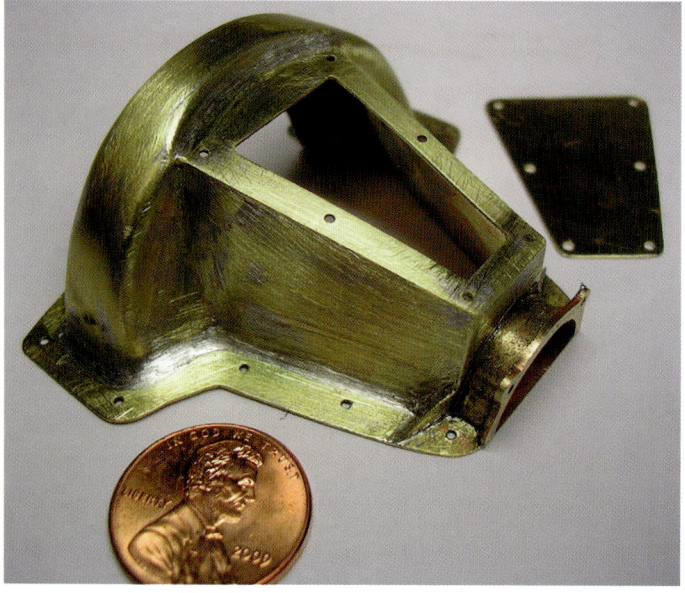

All tubes were cut on the marks. The tubes were then soldered in place from inside the cover. Once cooled, the excess flux was cleaned away with acetone on a Q-tip prior to sawing out the center section of the tubes by inserting the jeweler's saw blade through the access hole. Cleaning prevents the flux from dulling the saw blade.

Keep this cutting technique in mind for future applications for sawing operations on tubes where concentricity is needed.

Here is a picture showing the tubes in their final position with rods, springs, and transmission bands in place. The foot pedals were located on the left side of the rods.

Here is a similar application without having the tube cut. Note that on the left side, a tapered tube was machined on a lathe, then slipped on the main rod and soldered in place for the desired configuration. The right end of the rod has had threads cut on it for the locating nut.

This picture shows a close-up of the aluminum painted transmission cover in place. Note how the radii have all blended in with the formed walls and surfaces.

TIP: After priming and painting a part with threads, always chase the threads with a tap to clean and avoid chipping the paint finish.

Here is a picture of the completed working engine in its brass finish prior to electroplating with the final painted aluminum upper transmission cover bolted in place. The transmission cover was scrubbed with a firm toothbrush using pumice and water, dried, primed, and painted aluminum with no additional sanding or filler required.

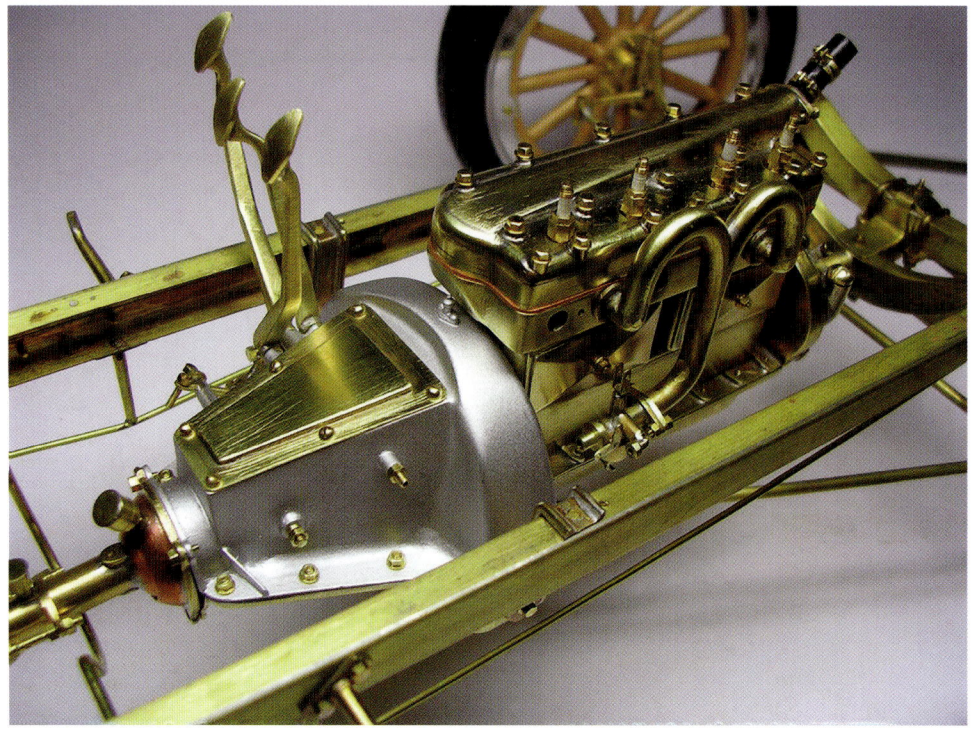

Part Fabrication: Various Stocks

Now, let us look at making some other part configurations using similar techniques. A good example of a hollow part would be the radiator for the Model T racer. It is based on a period radiator design used for racing. Modeling this required working with square tube, rod, bar, and flat sheet to achieve the end result.

Here is a picture of the real Model T race car showing the radiator to be fabricated. It was called the "arrowhead design." The design was an early attempt at streamlining, in addition to adding extra coolant capacity.

The first step was to develop a card stock mockup of the correct size and geometry, then test fit the mockup to the model chassis.

Once satisfied with the overall configuration, time was spent thinking through and deciding which parts would be made using which brass stocks. Scratch model building, unlike kits with step by step instructions, is at times like playing chess and having to think a few plays ahead.

The cardstock pattern also determined the inside configuration. Since the cooling fan would fit inside the final structure of the radiator, the thickness was important this time.

Card or certificate paper stock is a very useful product for making patterns of parts. It has enough stiffness to use as a pattern to trace onto brass sheet using a sharp scribe.

The back wall of the radiator was cut from .010" brass stock. The bottom core tank was cut and bent from one piece of .25" square tube. A "V" notch was cut on the center line of the tube length and bent to form the angle joint in the center using a drawn card stock template. By leaving the front wall of the square tube uncut, it formed the small radius needed when bent. The joint was soldered, then, moving down the sides, two more notches were cut, bent to conform to the template and soldered. All joints were cleaned of excess solder and sanded smooth. A 00-90 washer was soldered on the bottom side of the center joint, and a clearance hole was drilled to provide a critical vent hole that is necessary when soldering a hollow part. If a hollow part is soldered with no vent hole, the hot air pressure created inside will expand and separate the parts. A big enough hollow part could collapse when the hot air inside cools. Care should be taken with the cleaning and draining of fluids from hollow parts.

The back wall was carpet taped to a steel block with enough clearance to solder the ends of the square tube to the flat surface. A small steel square was used to assure a true 90-degree angle once soldered.

The chassis mounting brackets cut from flat bar were soldered at the top edge of the square tube.

A flat piece of .010" sheet was cut to the perimeter of the shape defined by the formed square tube and soldered while on the steel block. The small steel square was used to assure a parallel surface to the lower square tube. This piece was the bottom of the upper header tank.

A .010" x .50" brass strip was bent and soldered around the back wall shape, making sure the top tank stayed parallel to the bottom.

The upper header tank will become a hollow tank, so the filler neck was located and a pilot hole was drilled for the needed vent hole to release hot air pressure before being soldered.

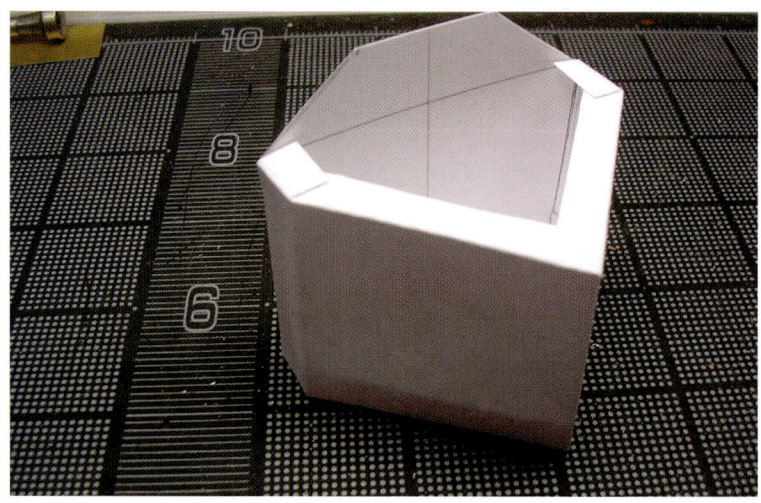

Sides glued to base unit.

Lower tank soldered to back wall of radiator, which is held to steel block with carpet tape to hold in place.

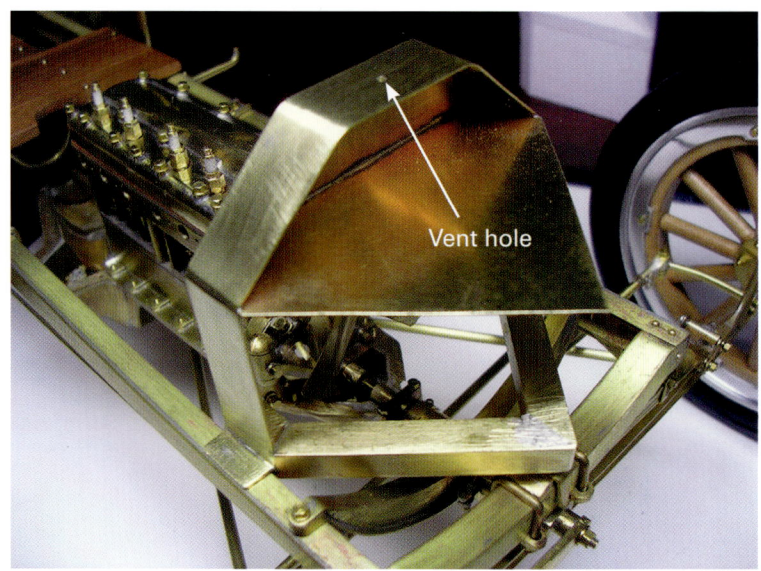

Sides and top of header tank added, note vent hole drilled before adding front to close header tank.

> **WARNING: Always drill a vent hole for hot air to escape when soldering a hollow form closed. If not, the part being soldered can blow apart under the pressure created by the heat.**

The radiator header tank top pattern was developed using card stock. Reference bend lines were marked while still taped in place. The double pencil lines help define the size of the radii to be bent.

Making card stock pattern for front of header tank

The top of the tank was formed, trimmed, fitted, and finally soldered in place. Excess solder was removed, and the whole assembly was cleaned with a Scotch-Brite® pad.

Now, the rest of the parts were added. The filler neck was made with two tubes and soldered in an enlarged hole that was the pilot vent hole. The coolant return tubes to the lower square tubes were formed using rod and soldered in place. A hole was drilled and filed elliptically to receive the upper water tube to the engine, and the joint soldered generously to form a larger radius. Finally, a .125" angle profile was notched and formed to provide a hood flange, then soldered in place. Here you can see the assembly in the raw soldered state prior to cleaning.

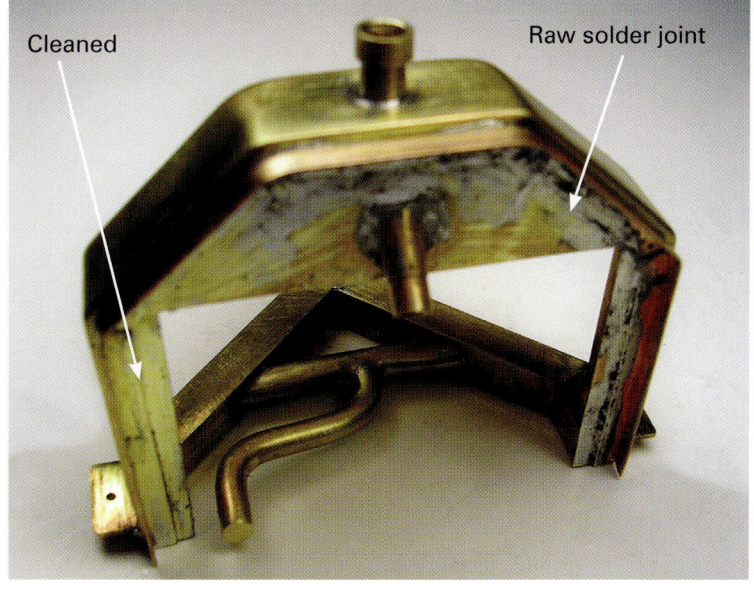

Hood flange soldered to radiator using 1/8" angle stock bent to contour.

The water tube was cut to length on the lathe almost all the way through and thus the rest of the tube became a handle to hold and locate at the correct angle while soldered. Once soldered in place, the tube was sawed off with a jeweler's saw using the lathe cut as a guide. The end was filed square and deburred on the inside of the tube.

Once completed and cleaned, the assembly was buffed with a small cloth buffing wheel and white polishing compound using a hand held rotary tool.

Here is a picture of the completed radiator with the cast resin hexagon core painted black in place. The resin core was why it was critical to have the upper and lower tank surfaces parallel to each other. The filler neck was threaded to receive the motometer.

Part Fabrication: Using Bar Stock

Brass bar stock, for the purpose of modeling, is available as solid flat bar, square bar, round bar, and hexagon bar. Bar stock is sold in many of the same sizes as tube of the same configuration. For example, flat bar (strips) can be purchased in various thicknesses from .015" x 2" x 36" to .125" x 2" x 36". Check suppliers for a complete listing of stock sizes available. An online supplier with an excellent assortment of brass is www.specialshapes.com. They do, however, have a minimum dollar amount per order, so plan ahead when placing an order. Explore their site for other products, tools, and materials. Another source of brass stock at local hobby shops are displays by K&S® Engineering. These outlets are good for occasional stock needs, but selection at times can be very limited.

In working with flat bar stock, it is best to determine the proper size of stock for the parts being made to optimize the yield of parts with the least amount of sawing possible. In the following example, I will share the making of functioning connecting rods that were used in the working Model T racer engine.

The configuration of the connecting rod was traced onto the flat bar using a sharp scribe.

The wrist pin and crankshaft holes were drilled first after centers were located using a small center punch to assure accuracy.

Note the center punch holes at the points of intersection of the scribe lines. These will be drilled to provide transition holes for the jeweler's saw blade. Transition holes are needed when changing direction to saw cut brass bar stock.

This picture shows the center punch drilled holes at the points of intersection of the scribe lines. This is when a drill press, cutting fluid, and sharp drill bits are essential. Care and patience must be used drilling such small holes with proper speed and feed so as not to break the drill bit.

Small drill bits are available in ten packs. Plan on acquiring packs of the most popular sizes for tapping and clearance holes. Start with these drill bit number sizes that correspond with scale hardware to satisfy most of your modeling needs: 48, 52, 53, 55, 61, 67, and 76.

Here you can see the four connecting rods drawn with a scribe and drilled on the flat bar ready to be sawed out.

The flat bar used here was .093" thick, .50" wide, and 12" long.

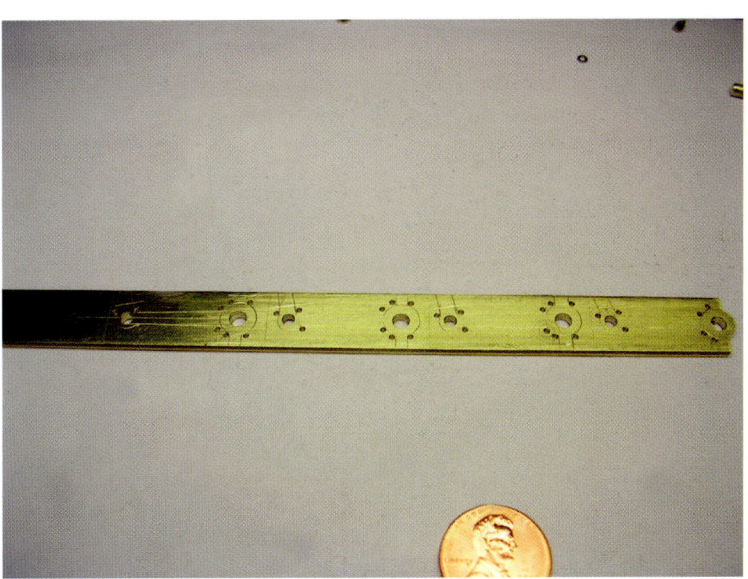

Carpet tape was applied to the top of the oak bench pin to help hold the part stable while sawing. Because doing this will help prevent the blade from catching and breaking, this is a good method to use while still getting used to using a jeweler's saw. This saw needs a certain "feel" to use effectively. The "feel" of using the saw is only learned with practice.

When sawing material thicker than sheet stock, use a thicker blade like a #1 size with plenty of beeswax as lubricant. Old, used candles can also be used for lubrication.

Here you can see the value of the transition holes. They enable shifting of the part and blade with little danger of breaking the blade while sawing.

TIP: When using a jeweler's saw to cut shapes out of flat stock, hold the saw blade square to the part and the part close to the edge of the bench pin. Use beeswax on the teeth of the blade as a lubricant and apply often while sawing. Remember to let the blade do the cutting. Excess forward force will only snap the blade.

Here is a close-up picture that better shows the saw cuts and transition holes in the flat bar.

TIP: When cutting out intricate parts, move the bar stock to keep the saw cut close to the edge of the bench pin. Think of the saw blade as a mini band saw that you move and adjust the part to be cut as you would with a band saw rather than turn the saw frame.

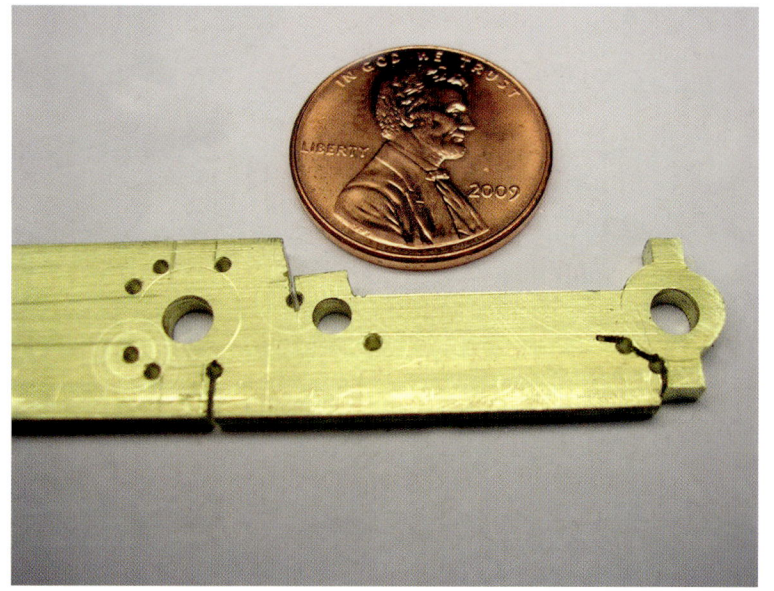

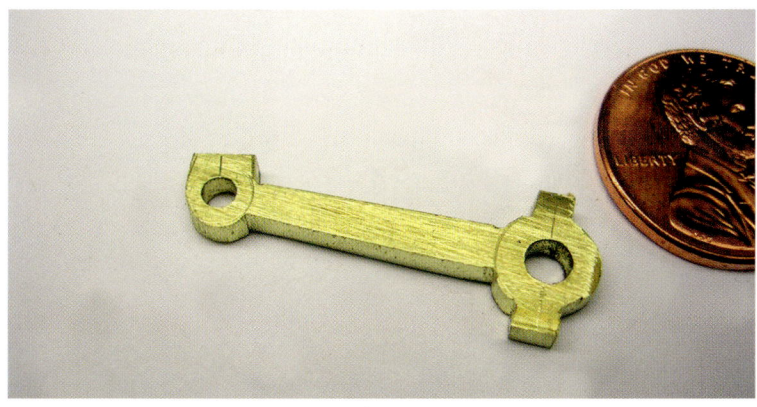

This shows the connecting rod perimeter cut out. Looking close, you can see the scribe lines for the subsequent saw cuts to and through the holes.

Here are all the connecting rods cut out. They were all slipped onto rods through the top and bottom holes to gang them together, then carefully filed so that all matched identically.

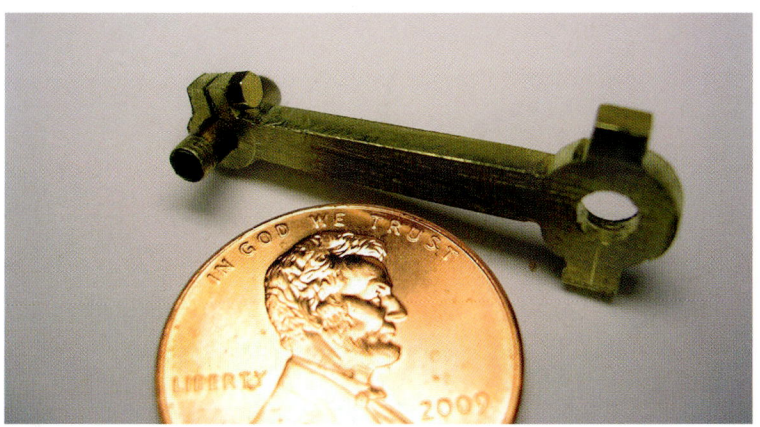

Once all matched identically, the wrist pin clamp bolt hole was drilled for a 00-90 bolt and tapped in each rod. Then, the saw cut was made to the wrist pin hole. A small piece of sheet brass was slipped into the saw cut, and the bolt clearance hole was hand drilled carefully. The small piece of brass prevented accidently drilling out the threads on the other half.

The bolt was then threaded in to assure it worked properly and held the wrist pin firmly. Use care to not over tighten the bolts and shear the bolt heads off.

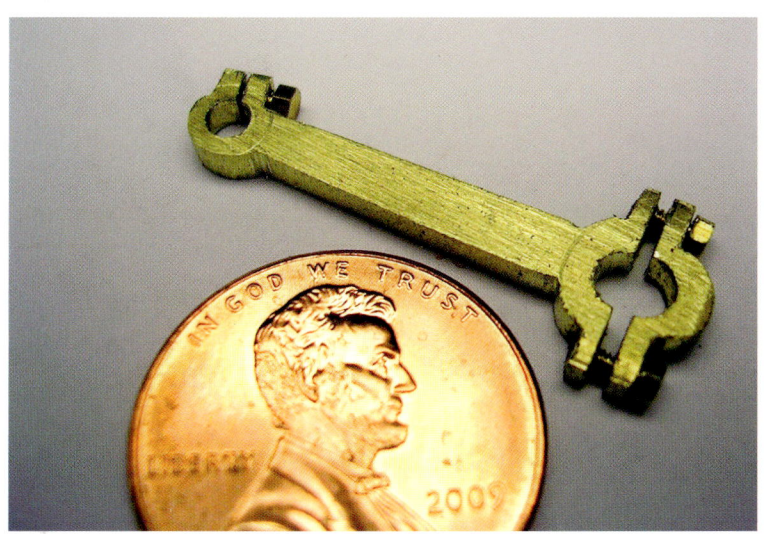

Holes were drilled for the crankshaft and sized for 00-90 bolts. Once all the holes were drilled, the outboard ends were filed half round using the hole as a reference guide. The holes were then tapped and following the scribed line, the clamp portion was then sawed off. Clearance bolt holes were then drilled on the lower clamp half.

Each rod was carpet taped to a carrier piece of aluminum taped to the bed of the milling machine. Using a small ball end mill, the relief slots were cut in both sides of the rod per the original connecting rods.

Note a locating hole was drilled into the aluminum carrier and a brass tube inserted to help locate the part in place while being milled.

This shows all the connecting rods completed and ready for installation. These are good examples of using flat bar to fabricate small, functional parts that have the strength to withstand hand operation.

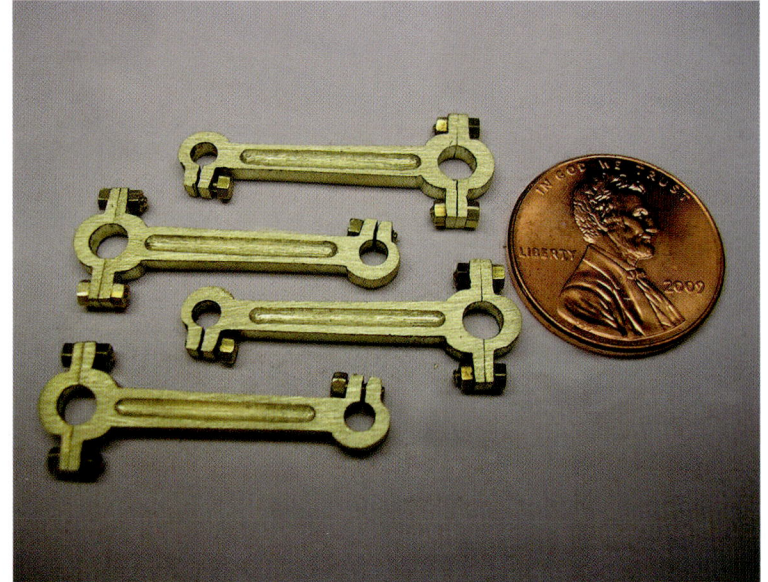

The connecting rods with wrist pins in the pistons are bolted in place. The wrist pins allow slight lateral movement of the pistons.

The piston assemblies were bolted to the crankshaft and ready to install in the engine.

The crankshaft assembly is shown here with rods and pistons test fitted in the engine block. Look closely and you can see the bolt heads holding the wrist pins inside the pistons.

The great advantage of model building with brass can be seen in this assembly. This assembly was fabricated by hand easier than using sophisticated, machining processes to build a small engine block with working components. The crankshaft end was chucked in the lathe and ran with no problems up to 5,000 RPMs. Just a touch of graphite powder was used as a lubricant.

Part Fabrication: Small Parts

In working with brass, one of the best advantages is strength of the material, which enables one to fabricate small, intricate parts that cannot be replicated with the strength needed in any other non-metal material. One of the great challenges in fabricating small parts is the ability to hold them while being worked on during the construction. The examples will describe the "Stick Method" to build small parts. This technique is referred to as the stick method because the part being built is on the end of a bar, rod or tube that became the handle to hold the part. The following first example is the fabrication of adjustable tie rod ends for the steering linkage of the Model T.

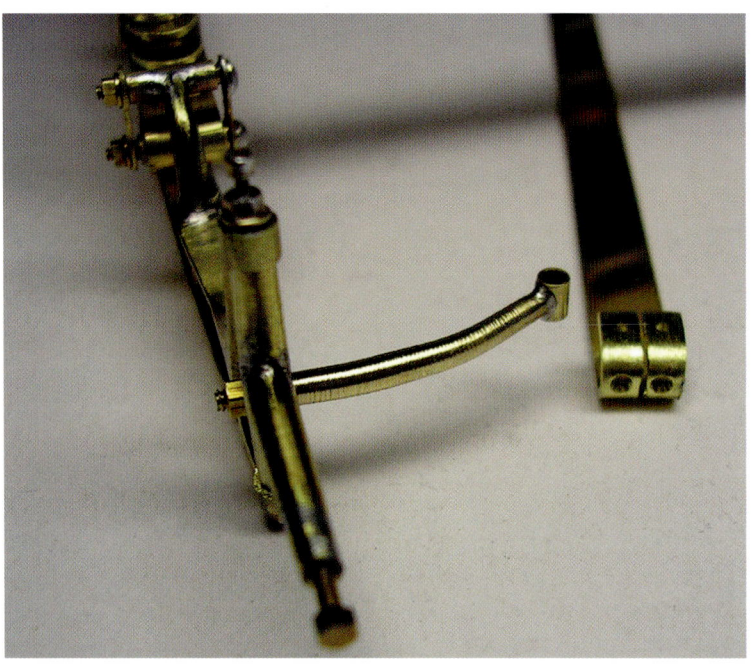

Form the drag rod yoke around a 1/8" drill bit shank as a mandrel. Drill two vertical holes with a #52 drill bit to receive 0-80 bolt. Drill two holes in the end with a #53 drill bit and tap for 1-72 threads. Once all holes are drilled, saw between them with a jeweler's saw back to meet the top end of the bend. Now there are two yokes started on the end of a 1/4" wide x .016" strip which will now be the handle to hold the parts.

To fabricate the tie rod ends, a flat bar the correct width and thickness was selected to form the end of the needed 180-degree radius. The vertical holes were located, marked, and drilled at the top surface, then continued through the bottom surface near the bend assuring perpendicular aligned holes using a #61 drill bit. The end hole was drilled and tapped to receive a 1-72 size bolt, which was soldered in place from inside the formed curve. The threading provided the mechanical interlock and strength needed for the linkage.

Then, very carefully with the jeweler's saw, the flat bar was cut down the center line equally on the top and bottom of the bar the needed distance for the cross cut to separate the parts once all the operations were completed.

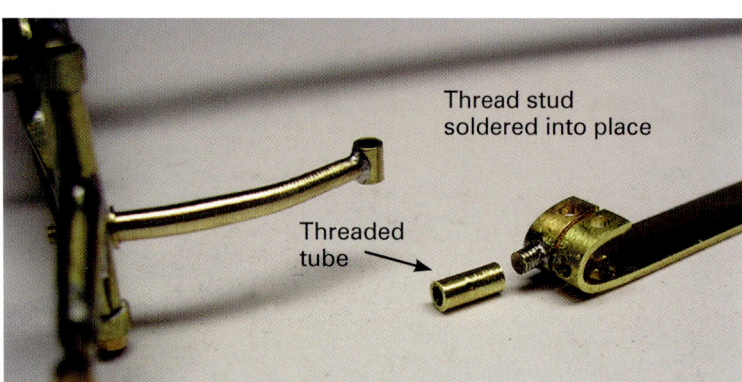

Screw 1-72 bolt into threads, paint threads with yellow ochre mix to prevent solder flowing up the threads. Solder bolt in place and saw off head. Or you can also use threaded rod. Tap 3/32" tube with threads to a depth of 1/4" and saw tube to a length of 3/16".

This will provide a very strong joint for the steering drag rod when the wheels are turned.

Yellow ochre powder can be purchased at an art supply outlet for mixing your own paint colors. Just mix powder with water and apply with a small brush. In effect you are painting on dirt and solder will not flow to it.

Here the 1-72 bolt can be seen. The hex head was sawed off after the bolt was soldered in place.

TIP: Before threading the bolt to the tie rod, a nut was screwed on and run up against the bolt head, then the bolt threaded into the end hole. Next, flux was applied from the inside. Threads were painted with yellow ochre from the outside curve to the nut. The yellow ochre prevented the solder from flowing up the threads.

Once the solder joint was completed, the nut was run down the stud, the bolt hex head was sawed off, and the end of the resulting stud filed square with a fine needle. The nut was then backed off, cleaning the threads after being dressed with the needle file.

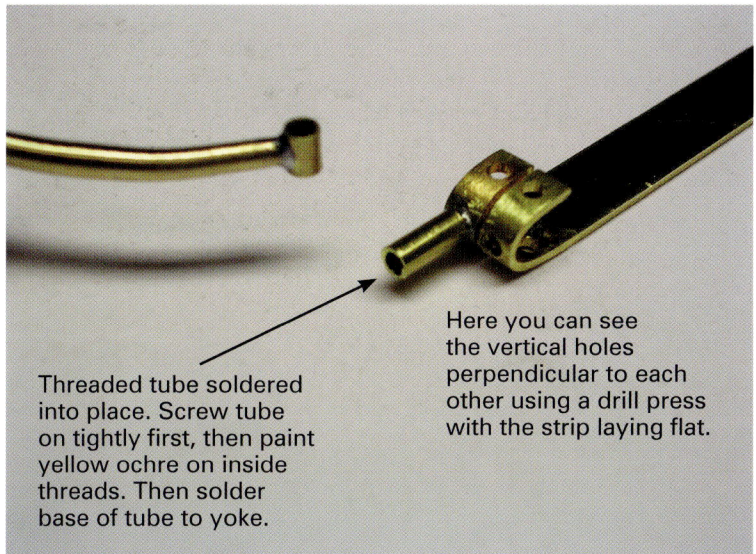

Threaded tube soldered into place. Screw tube on tightly first, then paint yellow ochre on inside threads. Then solder base of tube to yoke.

Here you can see the vertical holes perpendicular to each other using a drill press with the strip laying flat.

Next, a .093" tube was threaded to a depth of .25" and sawed off at a length of .187" with the sawed end being filed square with a fine needle file. The tube was threaded tightly onto the stud, and the inside threads painted with yellow ochre to prevent the solder from flowing up the threads when being soldered. Once ready, the solder was applied with a very hot iron and just enough solder was used to get a good, strong solder joint.

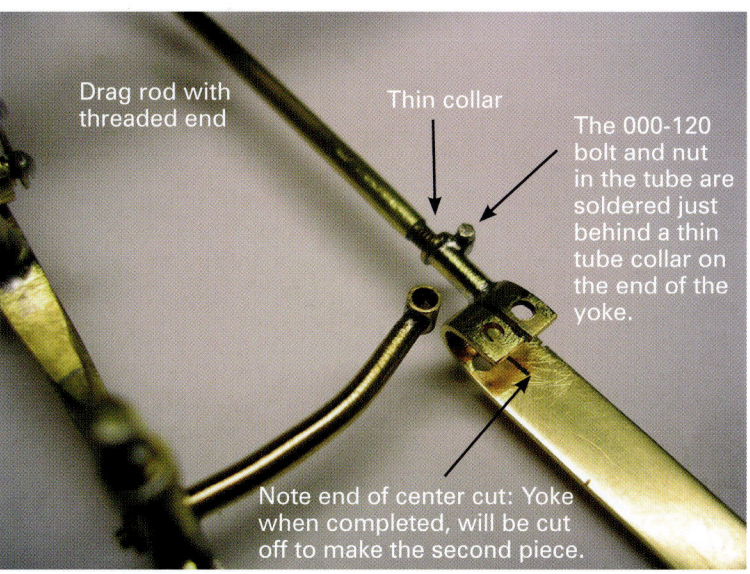

Drag rod with threaded end

Thin collar

The 000-120 bolt and nut in the tube are soldered just behind a thin tube collar on the end of the yoke.

Note end of center cut: Yoke when completed, will be cut off to make the second piece.

A locking collar was then formed and threaded with 000-120 threads. It was slipped on the end of the threaded tube and soldered in place. Once soldered and cleaned, the locking collar threads were cleaned with the tap. The collar was cut lengthwise with a fine saw blade #000 just enough to allow the clamping bolt to work. Plenty of cutting fluid was used to chase and clean the threads. Take care to prevent any residue flux from dulling the cutting threads of the tap. At this point, the top hole of the formed radius was drilled with a #55 drill bit for the 00-90 bolt clearance hole, and the bottom hole tapped for the 00-90 bolt threads.

Close-up of yoke before being cut off strip.

Here is the first finished tie rod end ready to be sawed off the flat bar. Test the fit of the threaded rod to assure the threads work while still attached to the flat bar. If necessary, the threads can be chased with the tap to clean them. Once satisfied that all parts worked, the first rod was cross cut off the flat bar and the ends filed half round with the 00-90 bolt and washer in place as a guide. The ends were filed half round while using the threaded rod as a handle to hold.

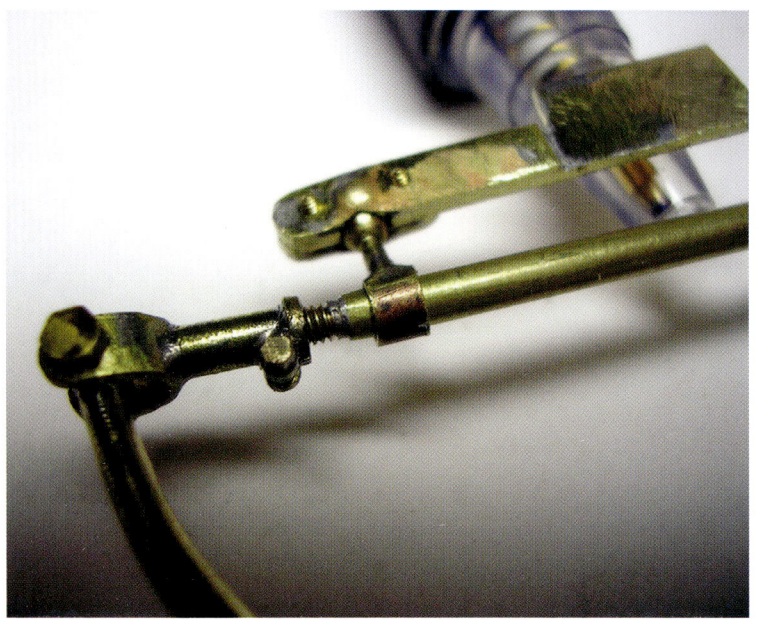

Here is the finished assembly. As per the real tie rod, the ends can be adjusted on the threads to set the front wheel angles, then clamped.

TIP: Care must be used when tightening small hardware, so as not to shear them off or strip the threads.

Close up of working ball and socket in progress. To make the socket, a piece of .016 was annealed by heating with flame until orange/red then quenched in water to soften. Drilled a 1/8" hole in the steel stock and then hammered into the hole with a 3/32" steel rod rounded on the end, working into a half round depression. Cut to size and filed one edge for ball shaft clearance. Then both were shimmed with center spacers with half round filed into them and then soldered to socket pieces. Drilled and tapped for 000-120 bolts. Once finally shaped, it will be cut off the handle piece and soldered to the rod.

Now to look at this variation of the "Stick Method" using round rod: Round brass rod lends itself perfectly to cutting and working on a tabletop lathe.

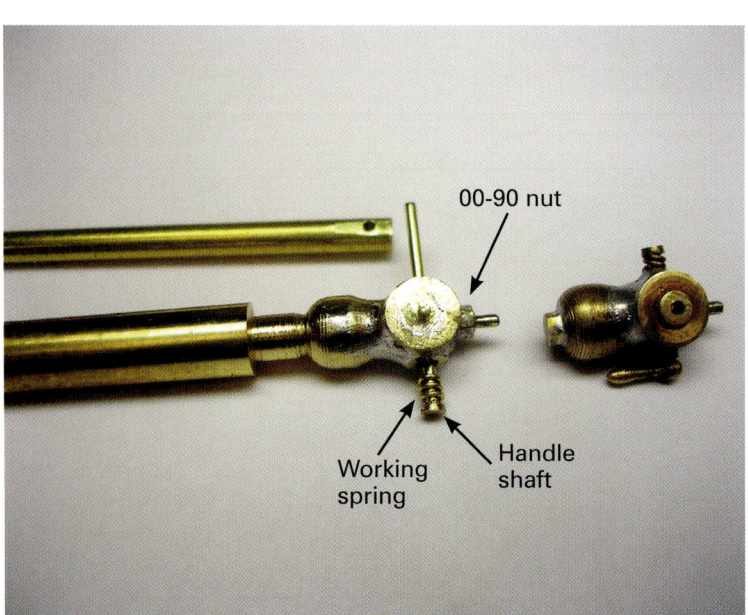

Here is the fabrication method of the fuel sediment bowls for the Model T gas tank. The bodies were turned on a lathe starting with a .312" solid rod. While still attached to the rod, holes for the handle and locating pin were drilled. In the locating hole, a turned disk with a pin was inserted into the hole and soldered in place with washers and nuts also being added to pins. All were then flood soldered to form a transition radius to the disk.

Sediment bowl turned on rod, cross drilled 1/32" washers soldered in place using toothpick to locate and hold while being soldered. 00-90 nut soldered to top and 2-56 washer soldered to face with pin.

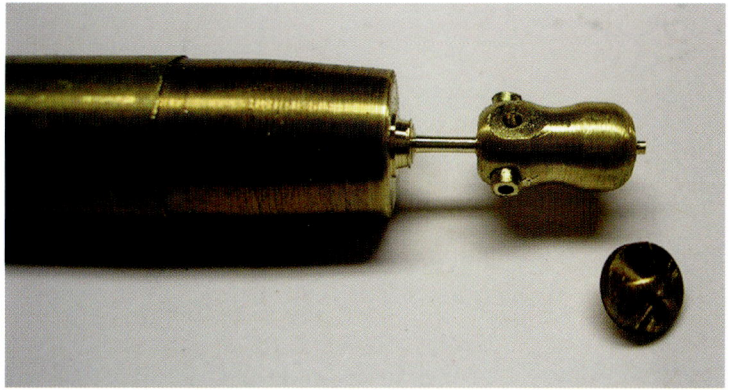

Here, the second finished bowl neck was turned down on the lathe to form the pin for the nut to be soldered to, then the stem was cut off with a jeweler's saw.

The lower part is the turned disk that was soldered in place before cutting off the rod.

Here is the sediment bowl shut off handle turned on the lathe. The handle shape was turned down to the hole area, then filed flat and the hole drilled. Once the hole was drilled, the turning of the handle shape continued down past the hole and polished with 400 grit sandpaper, then sawed off the rod.

In this picture, a .032" rod had a cap soldered on the end of the rod. A spring was formed around it. The handle was soldered to the rod with a little spring tension to hold the handle in place when turned.

Here is the finished, cleaned up, and polished fuel sediment bowl assembly with a shut off handle that turns. The overall height of this bowl is .43" high. The brass spring has enough tension to hold the handle in any position.

This shows one of the finished bowls in place on the gas tank. These bowls were copper plated and brass plated for their final finish.

This "Stick Method" example is a carburetor, which is a little more complicated to fabricate. With a lot of planning ahead and patience, fairly complicated small parts can be made of brass showing a great amount of detail.

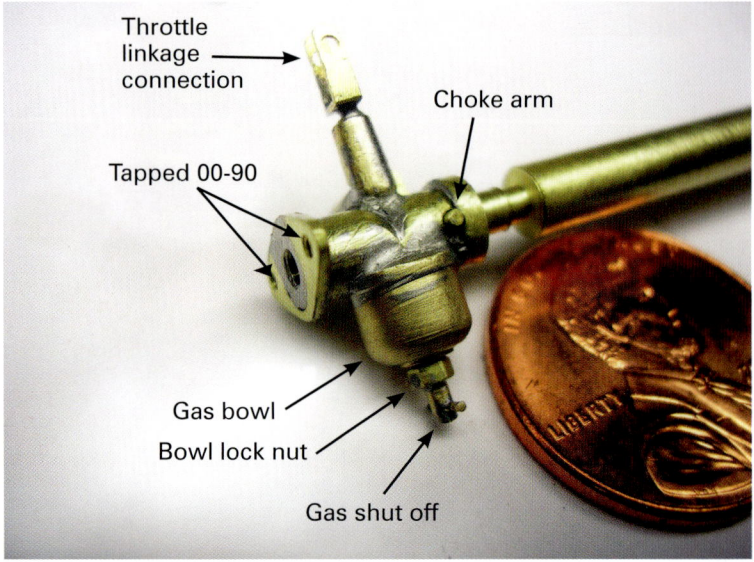

Finished left view.

This is a view of the carburetor bolted into position on the intake manifold. This is an example of a multiple-step plating process on one assembly. The entire assembly was nickel plated first, and then just the bowl portion was dipped into the copper solution and copper plated. Finally, the nut and shut off handle on the bottom of the bowl was nickel plated.

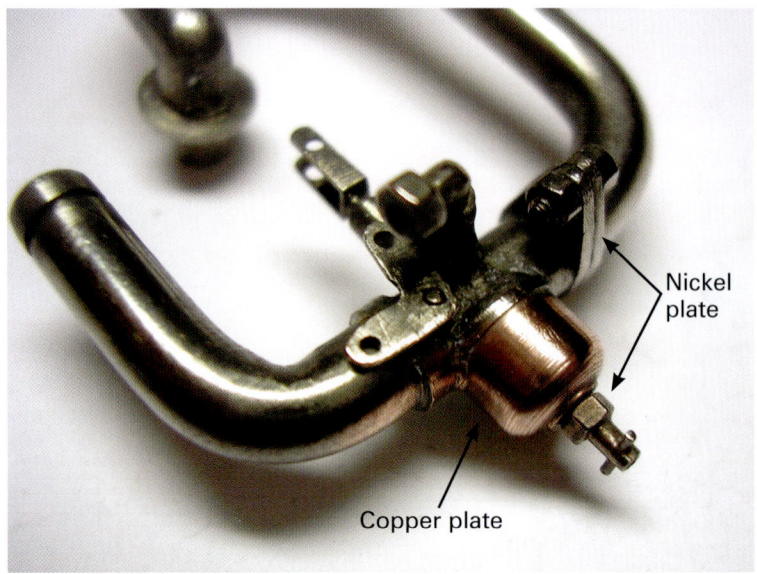

Dual plating of the same part. Carburetor bowl was dipped in copper plating and polished, then the drain nut was dipped in nickel plate.

This example is the first step of turning down the shape for a hand brake on a lathe. The handle end is held in place during turning with a live center. I will not attempt to teach one to be a machinist. There are great books available that can do a better job than I could. To learn about tabletop machining, I recommend **"Tabletop Machining"** by Joe Martin.

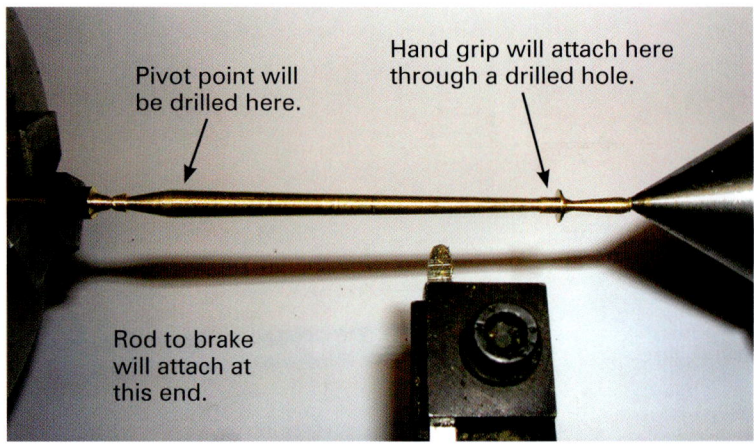

Rough turning for handbrake, this shape will then be filed into flattened oval of the real handbrake.

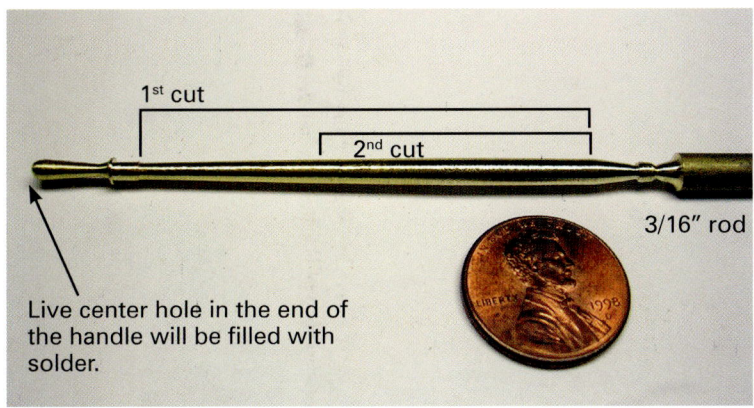

First taper cut from handle to pivot point was at 2 degrees; then from center to pivot point was 1 degree for final rough. Leave handle attached to parent 3/16" rod.

Here is the finished turned hand brake handle still attached to the "stick" rod. The rod acts as a handle to grip the part for subsequent operations, such as cross drilling of holes.

TIP: When turning any parts or components, always leave them attached to the "stick" rod until all needed operations have been completed, including cleaning and polishing. The very last thing to do is to cut the part from the rod or "stick."

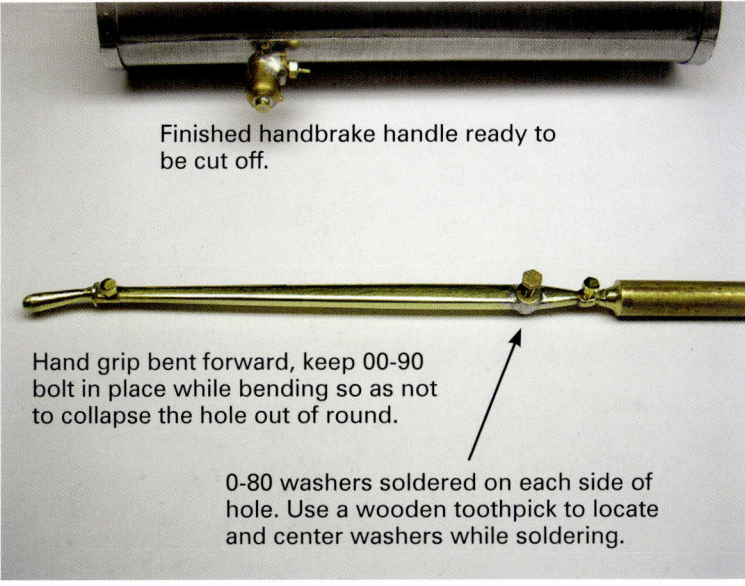

Here the turned handle has had the sides filed flat, and the needed holes drilled and tapped. The handle end was carefully bent forward to the correct angle with pliers.

Here is a close-up view of the functioning brake handle. This shows the handle shoulder was turned on the part, then the section below the shoulder was filed flat.

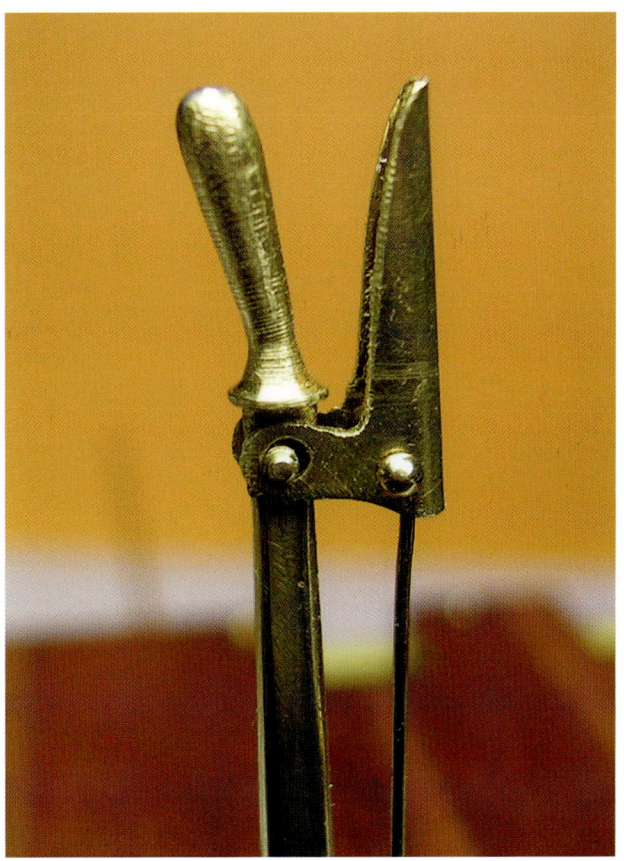

This is a view of the lower pivot area of the handbrake handle. You can see the ratchet pawl just above the pivot hole that has had washers soldered on each side of the pivot hole to form the shoulder for the locking nut.

Ratchet pawl is held in place with a 00-90 round headed screw.

Here is a view of the functioning hand brake handle in place. The .010" rod from the handle to the pawl actually works, releasing the hand brake as per the original.

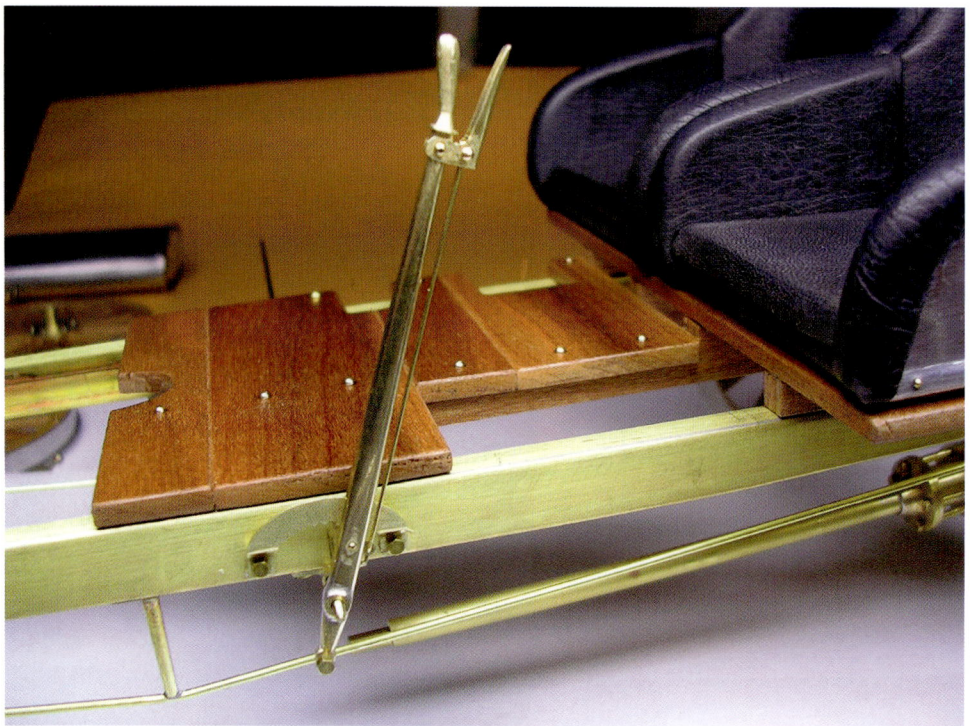

Part Fabrication: Using Laminations

There are times that parts and assemblies can only be fabricated using multiple laminated layers of brass stock soldered together. This is a great alternative technique to complicated machining processes. This process does require planning ahead and thinking through how to proceed and build the desired configuration. The lamination technique has been used to build parts for master patterns and to make molds to cast resin and metal parts.

Here is an example of building an engine block using the lamination technique. In the upper section of the pictures, .125" thick pieces of brass bar have been cut to the length and width of the engine. The internal areas have been cut out to reduce the overall weight. The pieces were clamped together using metal spring clamps and threaded brass hardware. These thick sections required the use of a propane torch to solder together.

The lowest section is the upper portion of the block, which consisted of both thick and thin brass pieces. The top piece had the cylinder locations drilled and the bolt holes located.

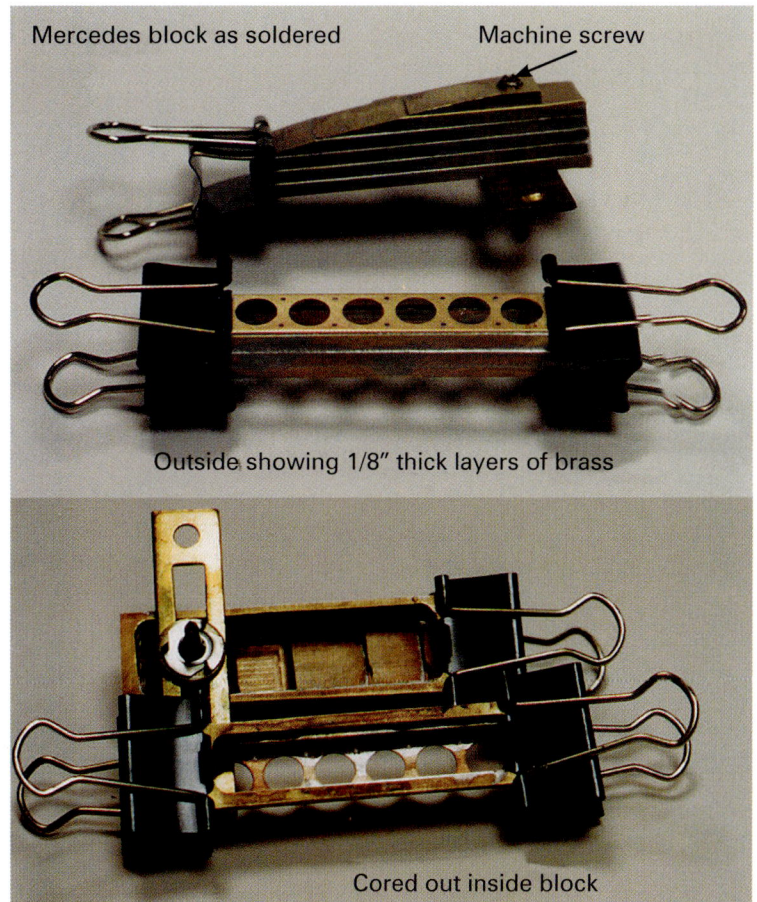

Mercedes block as soldered — Machine screw

Outside showing 1/8" thick layers of brass

Cored out inside block

This picture shows the finished block. The block was hand filed to shape, then the mating flange pieces and bottom fins added. The soldered layer lines can barely be seen, and, if painted, they would not be seen at all.

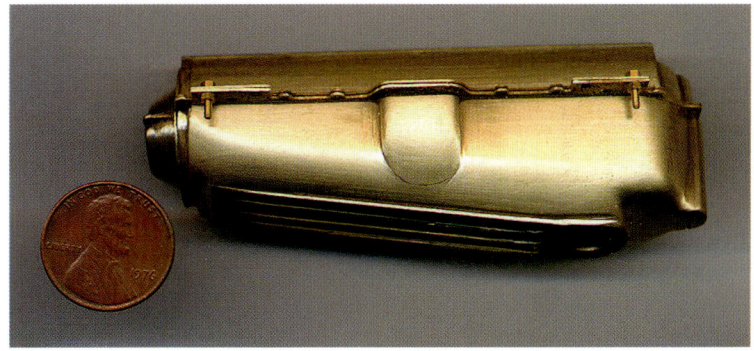

This top view picture of the upper crankcase shows the mating flange of which two identical parts were cut and drilled while spray glued together. The perimeter was cut using a jeweler's saw.

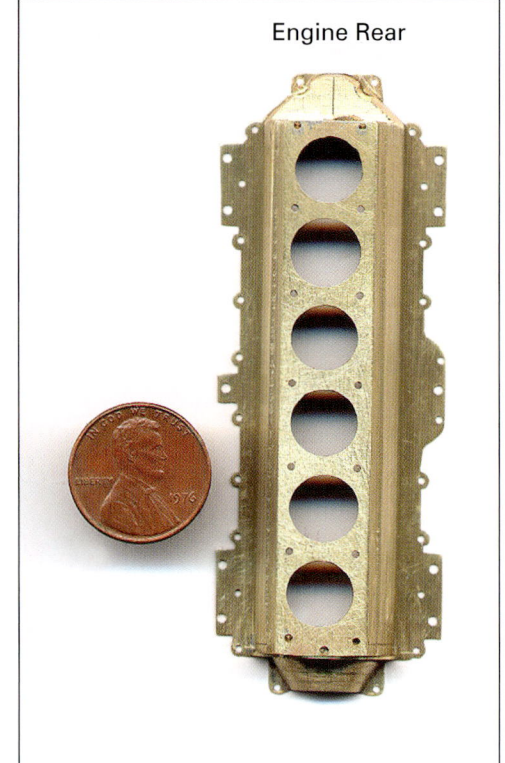

Engine Rear

Mercedes 180 HP 1/16th Scale Hand Fabricated Brass Cylinder Head.

This is the finished 1/16th scale engine completed and showing the small details that can be achieved using brass as the model building material. This engine was built using both hand fabrication as described above for the crankcases and a tabletop lathe to turn the cylinders.

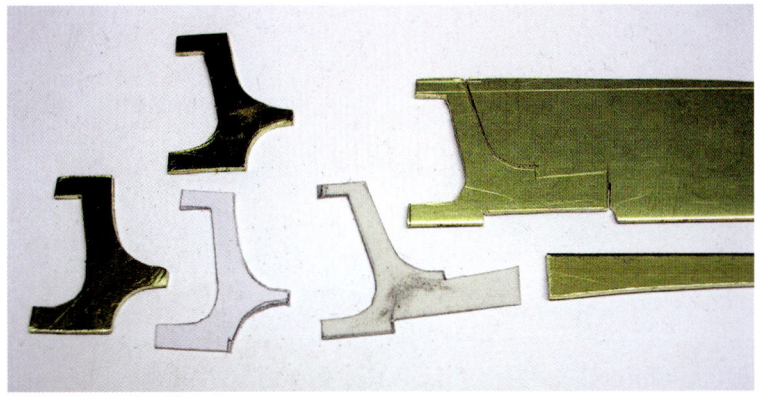

Laying out and sawing out pieces for spindle ends of the axle. The two that are already cut out are for the outside, while the piece partially cut out will be the center piece of a three layer sandwich lamination.

Some model building materials just do not have much structural integrity, so they are thickened beyond what would be the accurate scale for the parts. Car or plane axles would be good examples. This next sequence of pictures describes the building of a fully-functional front axle using the lamination technique. Here three parts were soldered together to form the spindle end of the axle. The center part had a longer tab that extended to where a tube would be inserted to have the shackle horn bolt through it.

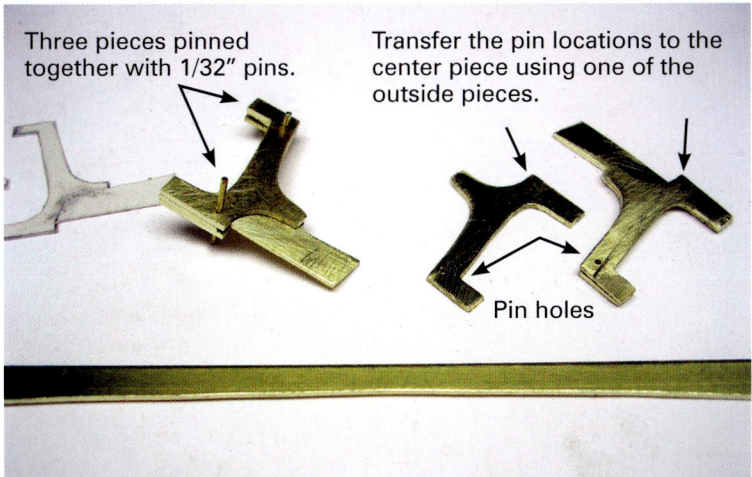

Three pieces pinned together with 1/32" pins.

Transfer the pin locations to the center piece using one of the outside pieces.

Pin holes

Once all the pieces were cut out, I Super Glued them together in a stack and then dressed them with a file to be identical. Then I drilled pin location holes while they were still glued together. Once apart, I then have four identical pieces with holes drilled in them. Transfer the hole locations to the center piece.

This shows the three parts to be laminated held together at the top and bottom with .032" rod pins. These pins located all three small parts together and in register to each other while being soldered to avoid shifting when hot. In this case, the parts were held with a metal spring clamp while solder was applied to the surface between the spindle horns. A hot iron flowed the solder to the opposite side with the tab. The pins were soldered in place, and these kept the parts in place during subsequent soldering operations. When cut off and filed smooth, the pins were not able to be seen.

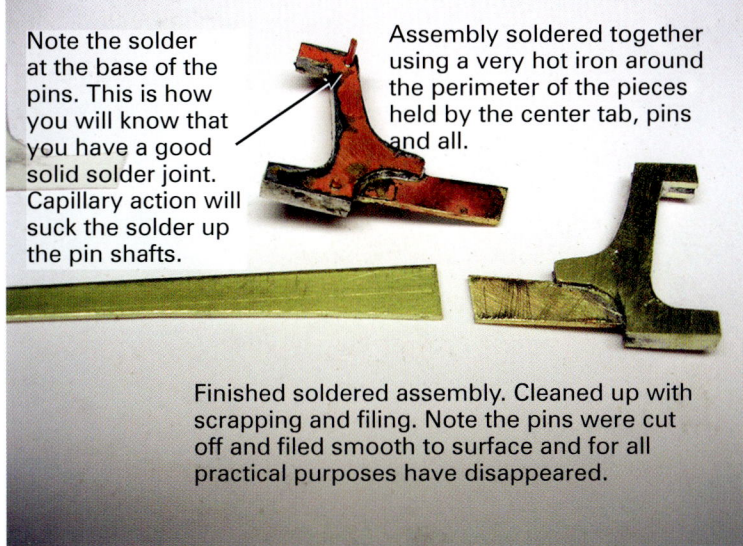

Note the solder at the base of the pins. This is how you will know that you have a good solid solder joint. Capillary action will suck the solder up the pin shafts.

Assembly soldered together using a very hot iron around the perimeter of the pieces held by the center tab, pins and all.

Finished soldered assembly. Cleaned up with scrapping and filing. Note the pins were cut off and filed smooth to surface and for all practical purposes have disappeared.

The use of pins is essential to hold the assembly together for subsequent soldering operations. Otherwise, they will fall apart.

Here you can see the soldered spindle assemblies. The one on the right has been cleaned of excess solder, pin ends cut off and edges filed to match. The one on the left is discolored and looks like copper. This is normal when using a very hot iron or flame. The copper in the brass rises to the surface. The brass color returns with cleaning and buffing with a Scotch-Brite® pad.

This shows the bottom flange, standing rib, and short tubes soldered in place. First, the bottom flange was soldered using small metal alligator clips to hold the spindle tab perpendicular to the flange. Then, the short tube was soldered in place, and a hole was drilled through the flange. Short aluminum tubes were inserted in the brass tube through the flange to keep the tubes in place while soldering the center section standing rib.

TIP: Solder does not stick to aluminum, so aluminum can be used to locate small parts during soldering.

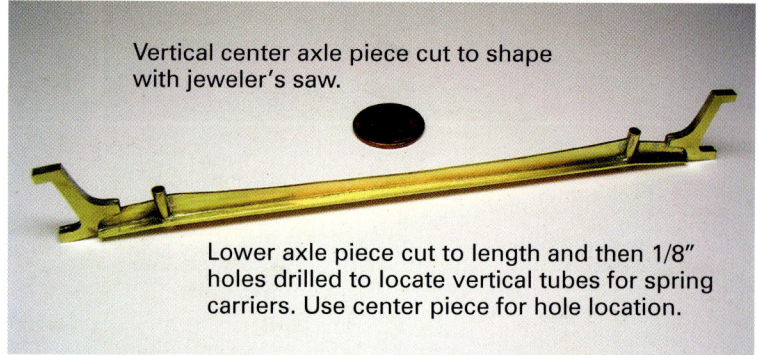

Vertical center axle piece cut to shape with jeweler's saw.

Lower axle piece cut to length and then 1/8" holes drilled to locate vertical tubes for spring carriers. Use center piece for hole location.

Once all pieces are cut and dry fitted, solder the center piece and tubes in place. Tubes should be a press fit to hold vertical and the center piece is held in the center with a spring clamp. Then locate and hold the spindle assemblies with alligator clip and solder in place. This is where the pins will hold the assembly together.

This close-up view shows the above soldering operation completed, and the assembly cleaned of excess solder. Note the center section profile was sawed prior to being soldered in place.

TIP: Always clean excess solder away prior to the next solder operation. This assures a tight fit of subsequent parts and gives good access before more parts are added.

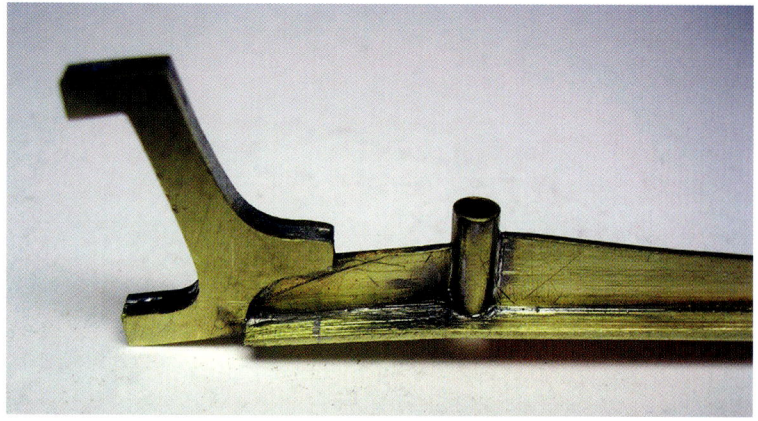

Close-up of semi-cleaned solder joint. Clean off excess solder before adding top piece and scrape with back of X-ACTO knife, engraving tool, and files. Note tight fitting joints.

Here you can see the top flange cut to the appropriate width and length, slightly bent and spring clamped in place to solder. Note the flange also was drilled for holes to press fit over the short tubes already in place. The flange should press fit into place with no tension or need of clamps to hold. The clamps hold the top and bottom flanges together during soldering. The 50/50 solder was applied to the backside in the center first with the clamps moved as needed to assure a good solid joint. The 50/50 solder was used in this case because of its lower melting point, therefore minimizing the risk to previous soldered joints being weakened.

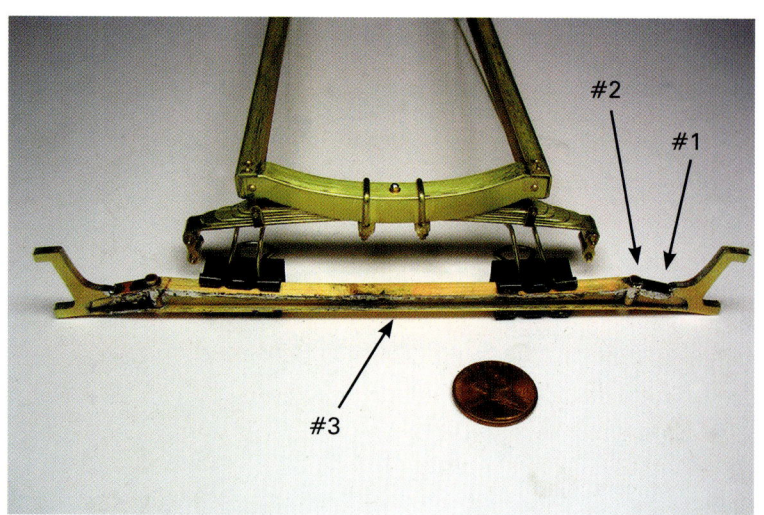

Top piece soldered in place. Using the big iron can make for ugly looking joints. Here, the two ends were soldered first. Start around the top of the tubes; then slowly down the center joint. Holding the whole piece vertically will help the liquid solder flow down (gravity). Note the amount of filing required.

Here is another close-up view of the above solder operation. The challenge is always getting the soldering iron tip into tight spaces. The solder was applied at the intersection of the tube and spindle end, then the tip was held on the top until the solder flowed into place. A very hot 120 watts iron was needed to accomplish this kind of soldering operation. This is an example of using the solder's characteristic of flowing to heat and placing the hot tip where the solder needed to flow.

The laminated front axle now has all excess solder cleaned off and is ready for final shaping by hand with assorted files.

TIP: Cleaning excess solder and flux will extend the life of files. Rub talc powder onto the file teeth. This will reduce the solder build up and act as a release when cleaning files with a file card.

Close up of raw solder joint. It looks a great deal worse than it actually is.

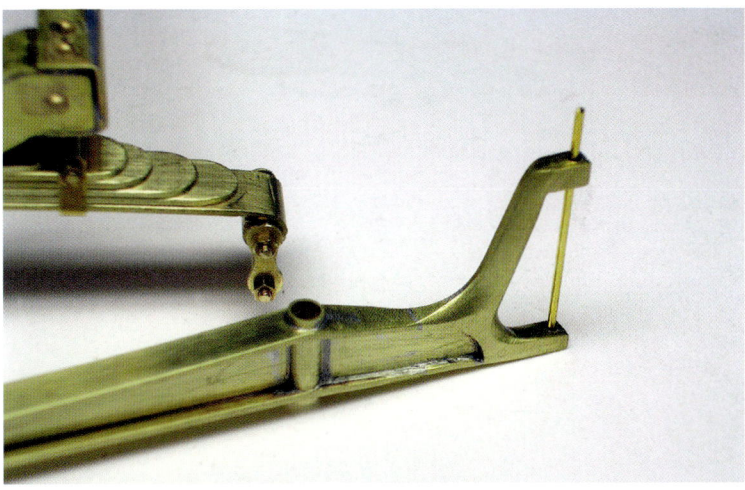

Whole assembly finished. All excess solder cleaned off. Spindle ends have been filed to required shape and holes drilled by hand. Use a small drill to start and work up to a larger size to obtain the correct alignment of the two. You will find that you spend more time cleaning solder off than the soldering operation takes. All shaping was done by hand with files only.

With the laminations done, this shows how the spindle end was filed to the final shape from squared parts that were started in layers. First, the front profile was filed using a half round file, then the taper toward the short tube was developed using assorted needle files. The holes were drilled while the ends were still square shaped. The ends were filed round using the holes as guides.

Close up of filed spindle end; will be final finished once wheel spindles have been made and final hole sizes determined. Note the vertical tubes both top and bottoms were filed down to correct sizes.

Here is the finished spindle end of the axle that functions as per the original. The spring horn was formed on the lathe with a taper, and the thin end had a locating pin turned on the end to locate the cross tube that was press fit onto the pin. Threads were cut on the other end to hold it in place with a nut at the bottom of the axle. The part was then slowly bent with pliers to the contour needed.

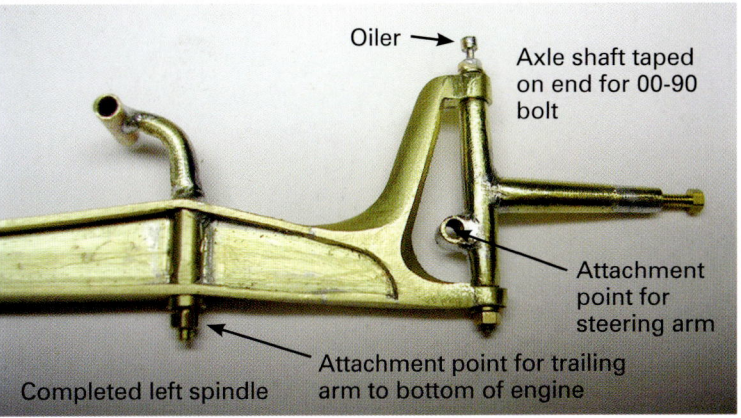

Oiler →
Axle shaft taped on end for 00-90 bolt
Attachment point for steering arm
Attachment point for trailing arm to bottom of engine
Completed left spindle

This shows the final axle assembly bolted to the chassis. Here you can see why the locating pin was needed to locate the cross tube spring horn and provide a mechanical interlock of parts with a tight press fit. The entire weight of the car will travel through this joint. Plastic or resin as a modeling material would not have the structural strength to support the weight of the finished model with these small parts molded at the correct scale.

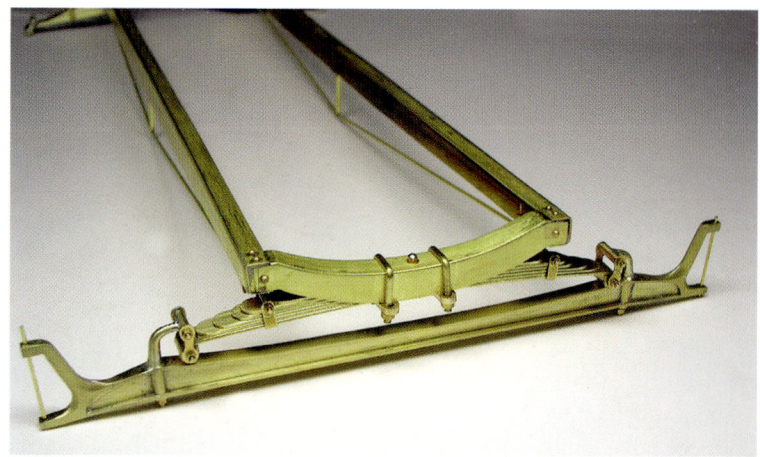

Axle spring hangers attached to spring assembly.

Here is a smaller example of a set of foot pedals that will be fabricated utilizing the lamination technique. The point of interest to learn here is the bending and forming of these parts. The first step was to scribe the outline of the part to a .032" sheet stock, then drill two holes. The hole on the right was for the pivot tube location, while the hole on the left was to make both a clean inside radius on the perimeter of the foot pedal and a transition hole when sawing with a jeweler's saw.

One side of the perimeter shape of the foot pedal was sawed with a jeweler's saw. This part became the master part to trace the rest of the parts needed. The pencil reference line in the middle was where the bend of the pedal was needed.

This shows the master part cut out with the pivot tube press fit in place. A good, tight press fit helped with the location of subsequent tracing of parts with a scribe and final positioning for soldering to match the three pedal angles.

A hole was drilled in the sheet stock, and the pivot tube inserted into both the master part and the sheet to be traced.

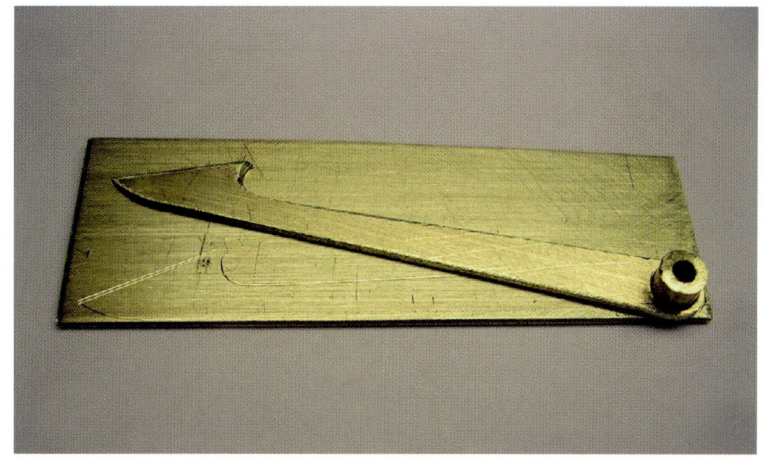

The pivot tube now locates the two .032" cut out pieces.

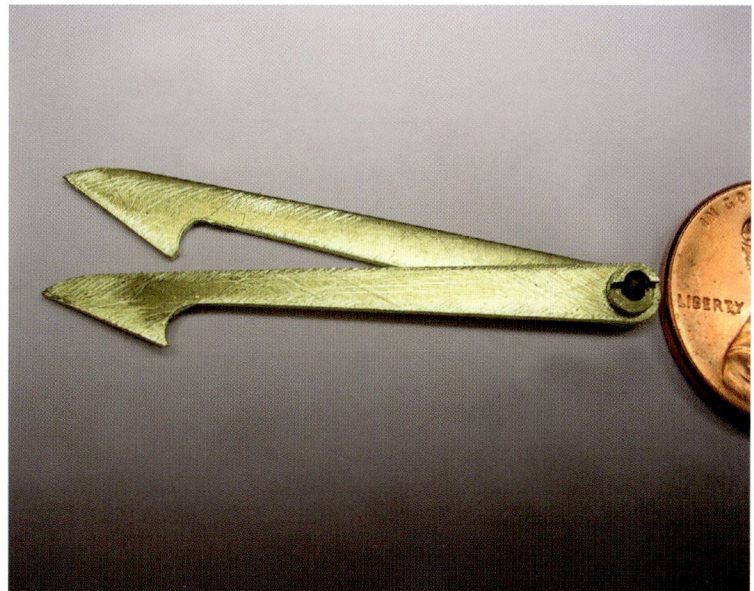

With both pieces aligned with the pivot tube, a .032" hole was drilled at the top near the inside radius below center of the flat portion, then pinned with a .032" rod. See the pin toward the top of the picture. The assembly was clamped with a spring clamp and soldered in place with a hot iron. The brass pivot tube was removed prior to soldering.

TIP: Use a short aluminum tube to align parts during soldering, then remove. Solder will not stick to aluminum.

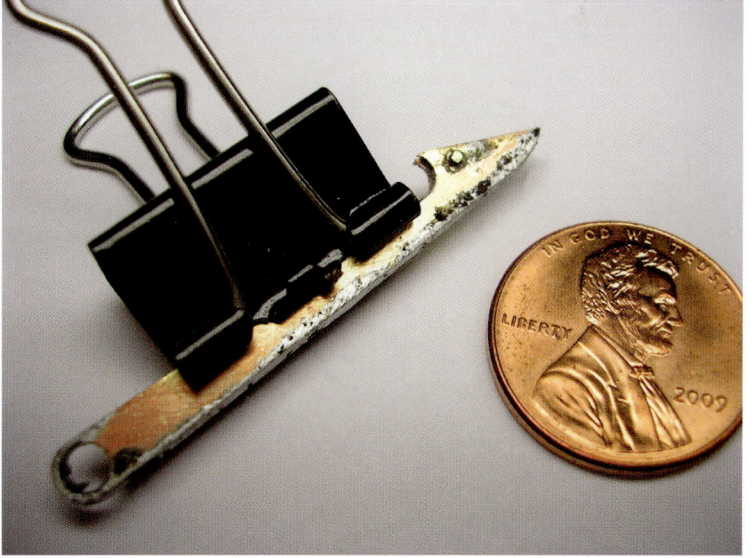

Once cleaned, the perimeter was filed smooth. A .032" hole was drilled through the flat foot pad area and a .032" pin was soldered in place to locate the foot pad. The pin was then trimmed and filed on the backside.

A third part for the lower half of the pedal was made only as long as to the bend point from the pivot hole. The part was then located to the soldered parts with an aluminum tube inserted in the pivot hole and soldered in place. This third shorter part was then filed to blend to the final shape prior to bending.

Two bending operations were preformed. First, the bottom with the pivot hole was clamped hard in a smooth jawed vise and struck with a hammer to the desired angle. The hole was then drilled back to round from distortion caused by bending. The upper portion of the pedal stem was clamped in the vise and hammered to the desired angle. The aluminum pivot tube was then press fit back into place, and the complete assembly soldered again, just in case any joint may have split during the bending. In this picture, the foot pad locating pin can be seen prior to trimming.

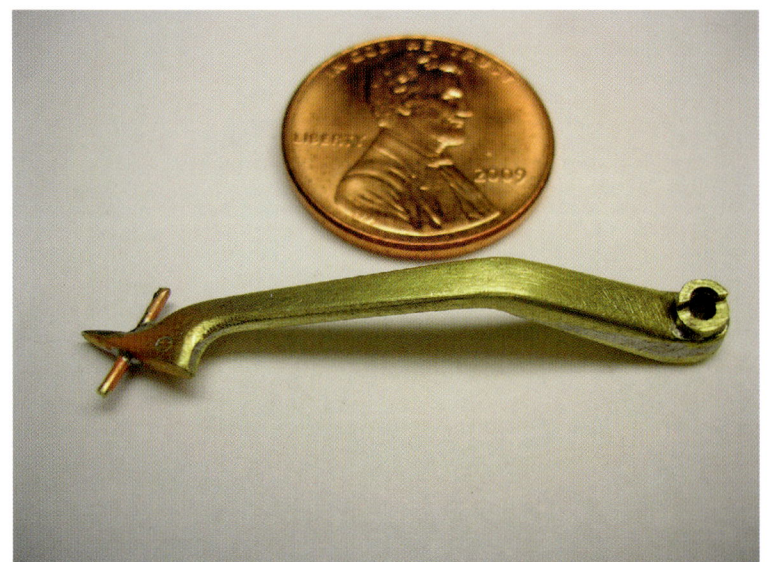

All three pedals are now in place. The value of the tight press fit of the pivot tube comes into play here, enabling the alignment of the pedals to the proper position. Once all three are in position, each one can be carefully removed and soldered in place.

Three identical foot pads were then cut out, filed true and a .032" hole drilled through the center of all three.

TIP: To drill small parts, stack them on top of each other using carpet tape or Super Glue to hold them together. The stacked parts are taped to a piece of hardwood, so all of them can be drilled at the same time to match.

This picture shows the final individual pedals, all formed as described to the different angles and configurations. Note the transitions from thick to thin at the bend points. Also note the backsides of the foot pads have had larger radii added with solder fill and filed to the final shape. These pedals were nickel plated for their final finish.

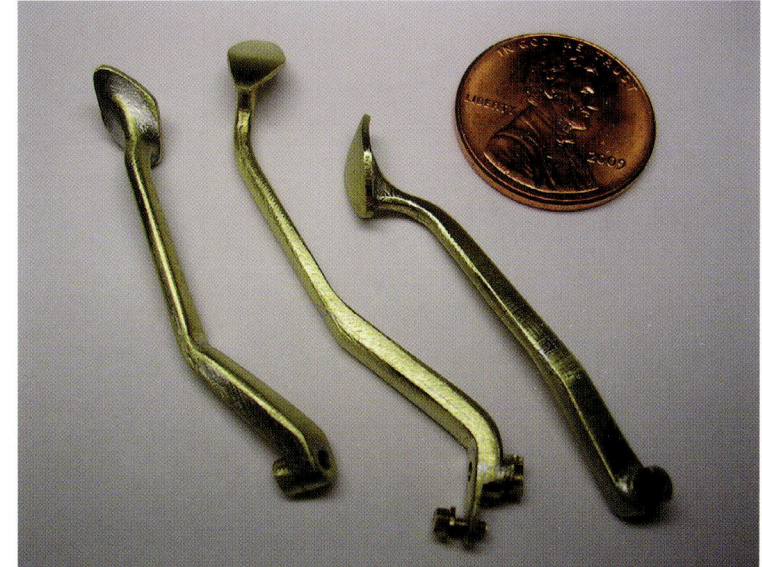

All pedals are in place and functional. They press forward and return to stops on the transmission tubes. Again, you can see the variation in thickness of the pedals as per the originals.

Using Binding Wire

There are times when parts of irregular configurations need to be tightly held together for soldering. One solution to accomplish this is the use of binding wire, a thin, very malleable steel wire available at most hardware stores. Short lengths can be cut and wrapped around the parts to be soldered. The ends can then be carefully twisted to tighten and hold the pieces together. Here a wood spacer can be seen being used to keep the sidewalls parallel with a spring clamp in place and being held with binding wire.

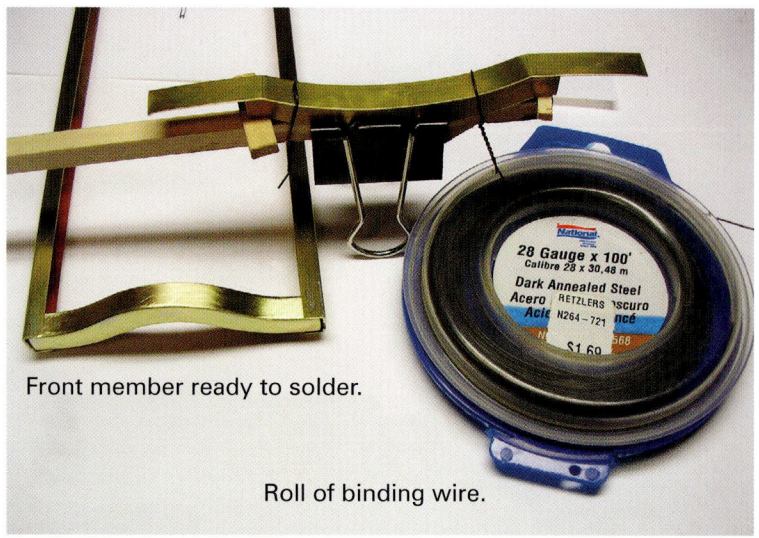

Front member ready to solder.

Roll of binding wire.

This is a very soft, malleable steel wire that works great to hold pieces tightly in place when soldering. This wire was purchased at the local hardware store.

This example shows the use of binding wire along with wood spacer blocks. The edges are soldered starting at the center and moving to the ends on both sides between the wires. The wires were cut away one at a time as the joint was soldered. Note one wood spacer is clamped into a bench vise to enable the use of both hands to do the soldering.

Engine mount ready to solder. Use wood pieces to maintain parallel spacing and not let the pieces collapse when tightened with binding wire. The top of the frame is hand formed to the contour.

Chapter 5
Incorporating Copper With Brass

Sometimes other metals are needed to form a part to obtain the desired geometry. One metal of choice is copper, which is inherently softer than brass, is readily available from the same suppliers as brass, and can be soldered to brass and electroplated if necessary. In sheet form, soft copper can be shaped into curved body panels as well. Brass in its half hard state will not work as well as copper. The following are examples of incorporating copper to achieve desired components.

Here a .020" sheet of copper has been rough cut and taped to the steel dapping block. Dapping blocks are tools used to form half round shapes and can be obtained in steel or hardwood.

On the left, the steel ball tool was first hammered carefully and slowly down the center to start the dome shape. The shape was then carefully worked in a circular pattern around the form to develop the entire dome. The hammer blows should be evenly spaced around the piece to be formed.

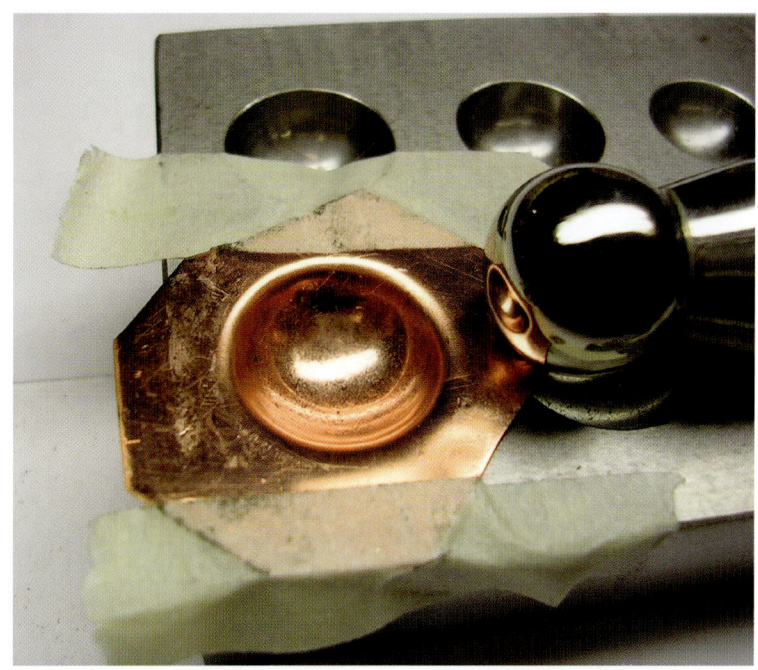

Once the dome was formed, a flat face hammer was used to carefully hammer the top surface to sharpen the edge at the perimeter prior to removing it from the block. This provided a flat reference surface for marking. In this case, the dome was sawed, and the cut edge filed, then placed back in the block. With soft hammer blows, it was trued up to eliminate any distortion during sawing. Also, by centering, you can use the block as a gauge to check your saw cut for evenness from the top edge of the part to the rim edge of the steel.

This is an overview of the parts and dapping block. If the flanges look familiar, these were made in Chapter 3 when making parts for the lower crankcase. Here you can see that the dapping block comes with various hole sizes. I used two different sizes to form the ball socket for the universal joint at the back of the transmission. The finished dome fits into the circle of the flange for a press fit joint that would then be soldered.

Here is a close-up view of the flange with the copper dome press fit in place. The flange has been drilled with the clearance holes for the 00-90 bolts.

The edge of the copper dome was filed to obtain the press fit. The flange hole was also filed at a slight angle to achieve the press fit.

The dome was soldered in place to the flange.

The flange was carpet taped to a wood block to hold the part during drilling. The center position was found by holding a flat file parallel to the surface of the block and just touching the top of the dome with it. This was done from three different angles to form the center flat spot. Then, the small pilot hole was drilled by hand on the center point marked with the tip of a sharp scribe in the center of the filed flat spot. The larger hole was then drilled using the pilot hole as a guide.

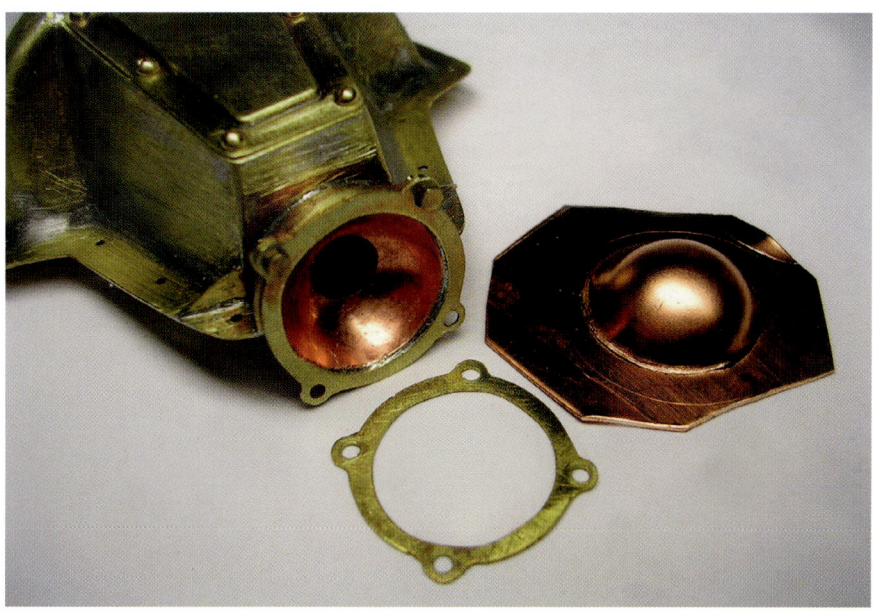

This shows the parts in various stages of completion. Note the sharp edge around the perimeter of the dome formed in the sheet. As mentioned earlier, this is accomplished while hammering the sheet on the top surface while in the dapping block. It also provides an edge to cut to with metal snips.

Part of the universal socket required the forming of a .75" sphere creating the ball part. This was accomplished by mating two formed spheres smaller than those of the socket halves. Here, they have been formed, checked for size inside the larger dome as seen in the previous picture, and taped together ready to solder.

A vent hole was drilled into one half on center to allow hot gases to escape during soldering. To hold tight during soldering, the taped together half spheres were clamped with spring-loaded tweezers to ensure a tight joint. Solder was applied to the exposed areas, then rotated 90 degrees, the tape removed and solder applied to complete the joint.

Once the joint was filed and cleaned to satisfaction, it was placed back into the dapping block. A small piece of carpet tape was put inside the forming dome to hold the sphere with the pilot hole pointing up.

The dapping block was used as a holding fixture while a clearance hole for the rod was drilled using a drill press. The sphere was then removed and reinserted in the block, and the center mark located. With small dividers, a cut-off line was scribed. The larger hole was drilled and filed down to the scribed line with a small bastard file.

This shows the ball filed and sanded smooth. Look closely and you may be able to see the solder joint. Here, you can also see the ball socket and how the flanges will trap it in place.

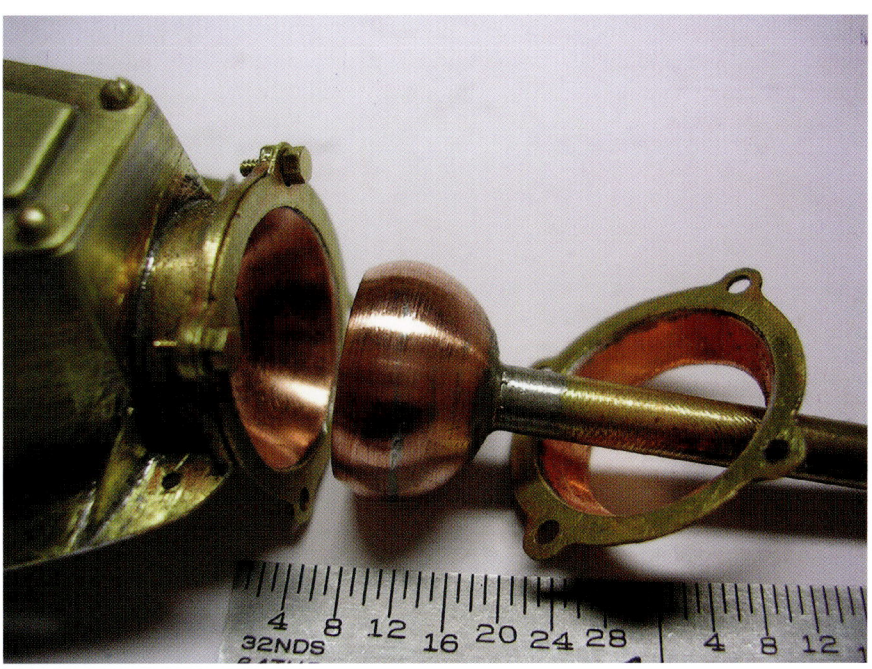

Here, the final assembly is bolted together with the ball socket in place and moving freely. The rod to the rear served as a handle during the cleaning and buffing operations. The rod would be cut to size once additional parts were added to it.

Here you can see the additional parts added to the shaft. The grease cups are threaded into the appropriate locations. On the socket half, a hole was drilled using a #61 drill bit for tapping 00-90 threads. A wood toothpick with the washer was located in the hole and soldered in place. Then, the threads were cut. The same was done on the shank with a shaped piece of brass set against the shoulder of the flange. The shank piece with the flange was turned on a lathe. The shank piece was then filed down to its final oval shape, slipped on the rod, and soldered in place.

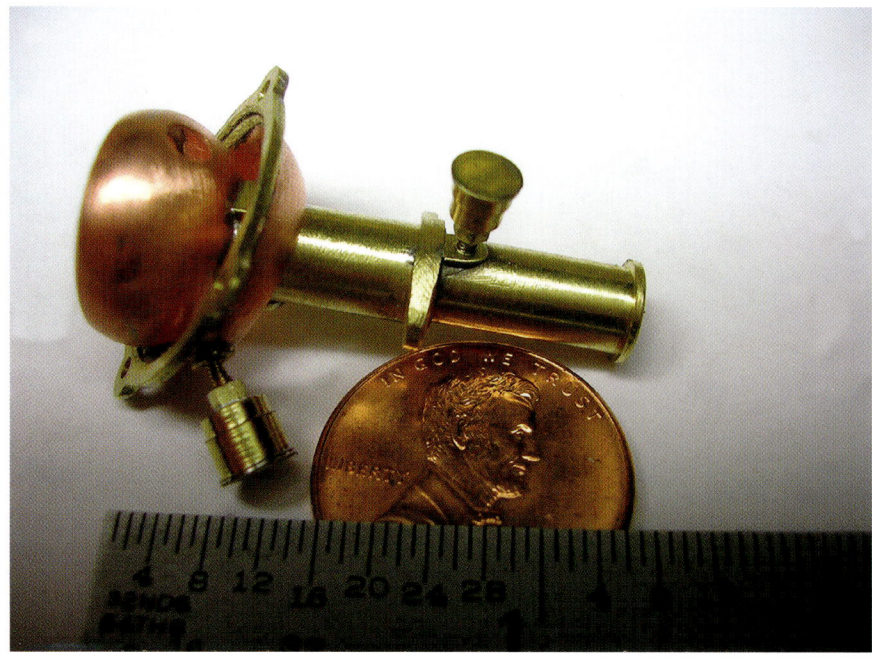

Here, the final assembly of socket and ball are in place. These parts then dictated the final size of the universal joint that had to be fabricated and fit into.

The grease cups are turned on a lathe starting with 00-90 bolts that short rods were soldered to the ends, turned, and threaded. The caps are threaded tube with the top soldered on and trued up on the lathe. The caps are removable.

The gas tank is another example of forming with copper. In this case, brass sheet stock could not easily be formed as one piece with both ends of the gas tank being rounded to smooth, compound curves. A carved wood pattern was made from a wood block cut the width and length minus the wall thickness of the copper sheet, which was .020" dead soft copper. The sheet of copper was centered on the top of the wood form and carefully hammered with a leather mallet to start the forming process. After the shape developed down the sides and ends, a steel hammer was used to slowly continue the forming.

Lower half of the gas tank being hammer formed over wood buck. Copper is more malleable than brass to form round ends with no seams. In progress, not finished.

Wood buck carved to half shape of gas tank. Two halves will be formed and joined with details added to each half.

Here you can see the block is deeper than the required gas tank shape. The copper form was annealed after the initial shape was developed. The copper was heated with a flame until very hot, then quickly quenched in water. This softens the copper and eliminates the work hardening from hammering. Continuing with gentle hammer blows, slowly and evenly, the final form was set against the wood pattern. The flat surface of the wood block then enabled a straight line to be scribed around the part to establish the cut line. The final smooth surface was obtained using files and sandpaper.

One half of the gas tank started. Will be formed to final shape and trimmed to final dimension.

Once satisfied with the finished surface, it was buffed and polished using a metal polishing compound. Then, all the finished detail parts were soldered in place.

TIP: When combining copper with brass parts, always keep in mind that the copper is much softer than the brass and is more easily scratched or damaged when cleaning off excess solder, so work carefully.

Finished gas tank hammered out of .20" soft copper to have one piece seamless.

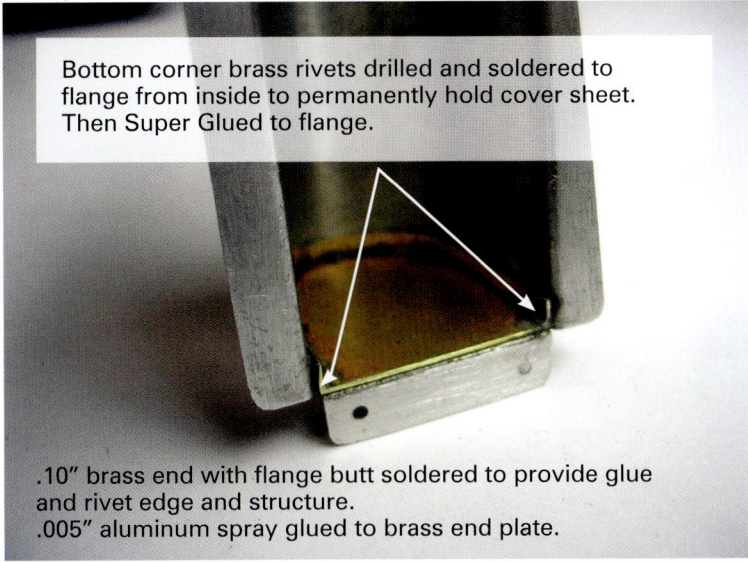

Bottom corner brass rivets drilled and soldered to flange from inside to permanently hold cover sheet. Then Super Glued to flange.

.10" brass end with flange butt soldered to provide glue and rivet edge and structure.
.005" aluminum spray glued to brass end plate.

USING OTHER MODELING MATERIALS

Sometimes modeling offers challenges to making certain parts. An aluminum cover was needed for the magneto box. The easy solution would have been to use glue. However, I wanted a more long term, permanent solution. Two identical brass end caps were made with a flange facing inboard enough to drill through for brass rivets. The end caps were then spray glued to the aluminum cover sheet with the brass end caps set in the thickness of the aluminum sheet. Holes were drilled at the lower corners and top centers for rivet locations. Brass rivets were soldered in place on the inside of the cover. The rivets penetrated through the aluminum sheet and brass flange, thus assuring a mechanical joint.

This shows the cover in place with the lower corner and top rivets in place and the aluminum end caps fitting flush to the edge of the cover.

TIP: Sometimes it is difficult to hold things in place until certain operations are completed. Clamps could not be used in this situation. In this case, the top holes were located first, and rivets were inserted. Super Glue was used to hold the aluminum cover edges to the brass end cap. The glue held the parts in place during the drilling and soldering of the brass rivets.

.005" aluminum roof flashing used to form cover and polished up a little before adding the rest of the brass 1.2mm rivets.

Using masking tape as rivet hole locator, holes drilled along the edge of the tape. Tape was then used at the other end.

Rivet locations were located using reference marks on a piece of masking tape indexed on the three rivets already in place. The same piece of tape was saved and used on the other end. Once all the rivets were done, the aluminum end cap edges were filed very carefully—first with a medium, then a fine needle file to match and round the cover edges. The aluminum edges were then carefully polished by rubbing and burnishing with the polished round steel shaft of a scribe.

Another example of combining metals is the brass firewall frame with the polished sheet aluminum firewall. The brass frame is the structural piece that is bolted to the chassis.

TIP: Avoid soldering operations around aluminum parts. The spatter from the flux will mar and damage the aluminum surface finish. Parts can be covered with masking tape to avoid this damage.

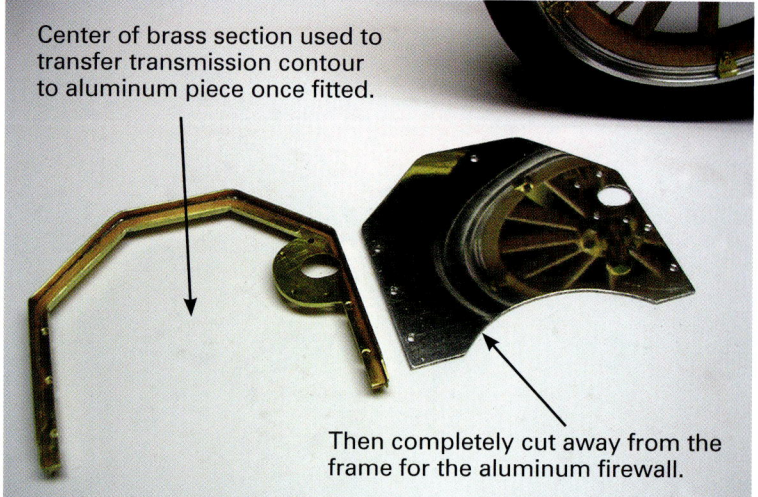

Center of brass section used to transfer transmission contour to aluminum piece once fitted.

Then completely cut away from the frame for the aluminum firewall.

The firewall is bolted in place to the chassis. Also notice the parts screwed and bolted into the wooden magneto box.

Wood can be tapped to accept screw and bolt threads. Ideally, a hardwood is used. Tap the threads in the hole and place a drop of Super Glue into the threaded hole. Let the Super Glue dry, then chase the threads with the tap to clean. The Super Glue hardens the wood fibers; however, care still must be used to not overtighten screws or bolts in the wood.

Commutator harness attached to coil terminals.

Leather is another material that can be incorporated into a brass model. The Model T hold down straps or seat belts are more realistic out of the intended material. This picture shows the various parts required to make the hood hold down straps.

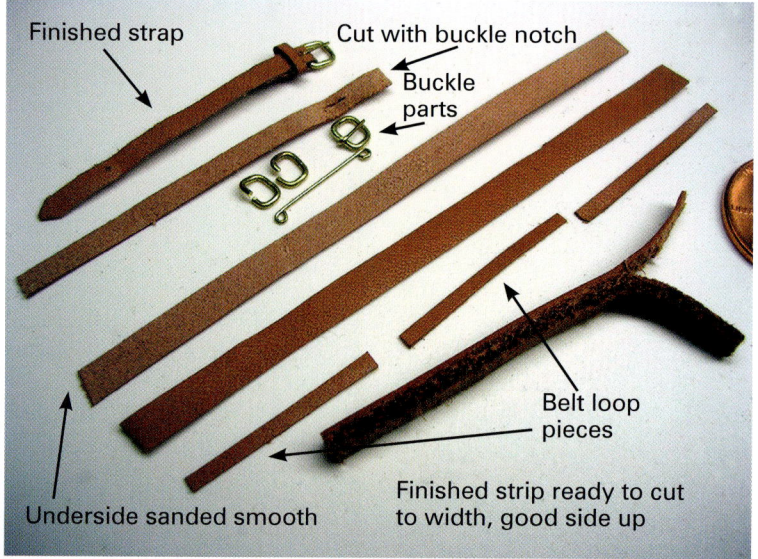

Finished strap

Cut with buckle notch

Buckle parts

Belt loop pieces

Underside sanded smooth

Finished strip ready to cut to width, good side up

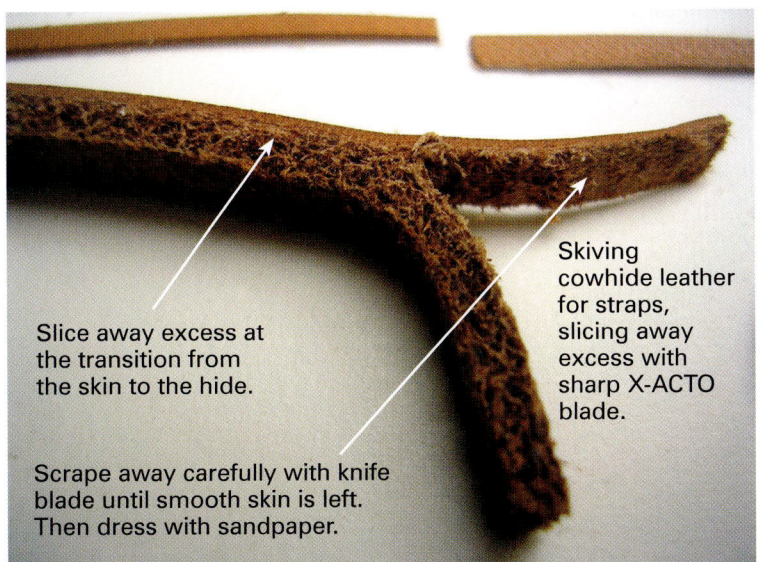

Slice away excess at the transition from the skin to the hide.

Skiving cowhide leather for straps, slicing away excess with sharp X-ACTO blade.

Scrape away carefully with knife blade until smooth skin is left. Then dress with sandpaper.

Taking cowhide down to the proper thickness is called skiving. Skiving cowhide leather is somewhat easy once you get the feel for it. Use a sharp X-ACTO® blade and trim away the excess from the outer skin, then carefully use sandpaper to smooth the back of the cut surface.

These belt buckles were formed from .032" rod using needle nose pliers. The indexing tongue rod was formed around a .032" drill bit as a mandrel and trimmed to fit snug on the buckle. The buckle ends were offset to enable the loop to slip on, be crimped together, then soldered.

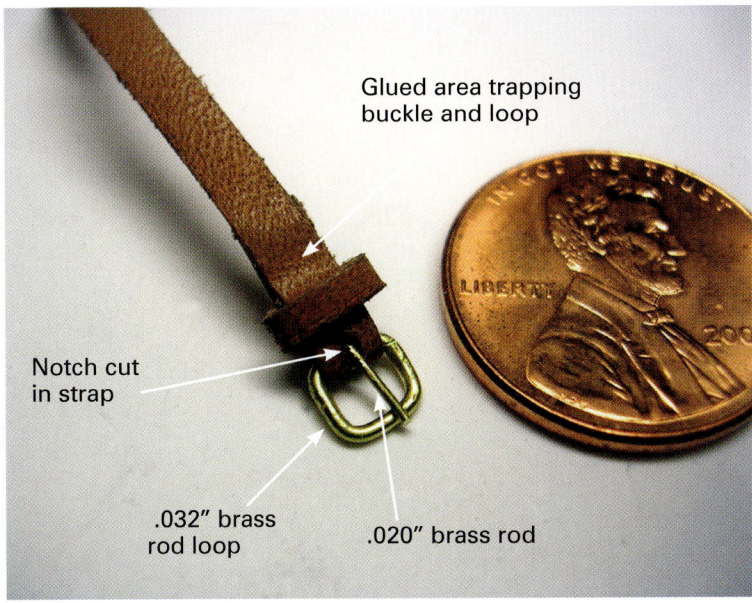

Glued area trapping buckle and loop

Notch cut in strap

.032" brass rod loop

.020" brass rod

This shows the completed belt buckle.

103

Here, the hold down strap is in place through a hole in the hood and looped around the chassis rail.

Another example of leather used in a model is this trunk carry handle. The leather handle is indexed through sections of rectangle tube soldered to the trunk frame.

Close-up of leather strap carry handle.

Making a Wire Wheel

Wire wheels have applications for airplanes, cars, and motorcycles. Here are the fundamental building steps to make them with purchased O-rings and brass. First, the rim was made to size starting with .010" strip of brass annealed and formed. In this case, a half round rim shape was formed between two strips of bar stock that were carpet taped to a piece of wood board. The brass strip was taped on center between the flat bar stocks. The half round shape was slowly formed with the end of a drill bit shank. The rim was hand formed to size, trimmed, and butt soldered using the O-ring as a fixture. Two .032" brass rod rings were then soldered to the outside edges of the formed rim while the rim was still affixed inside the O-ring. The O-ring resists the soldering temperature. Spoke holes were located and drilled for .020" spokes.

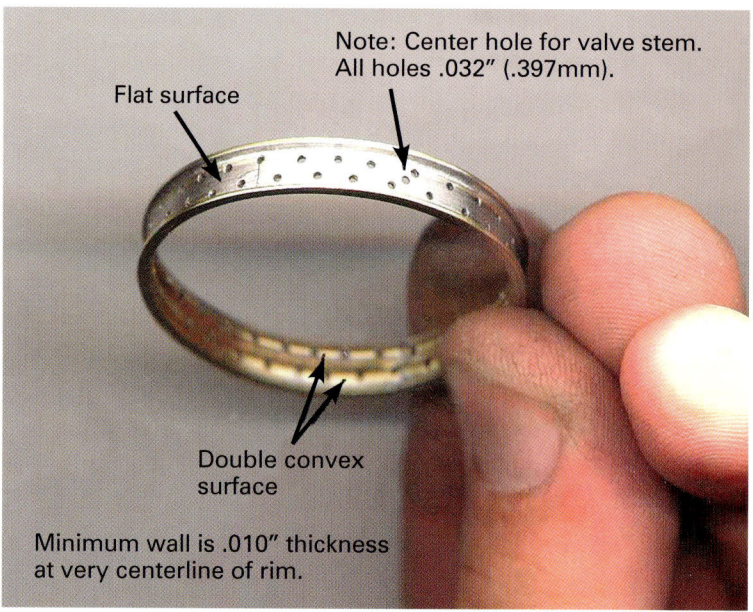

Sopwith Wheel Rim - Master Pattern.

Here is the finished 1/15th scale wire wheel for a Sopwith Camel. It required sixty-four .020" brass rod spokes.

Also pictured is a typical spoke assembly fixture. The hub was turned on a lathe to assure concentricity, and each flange was drilled with thirty-two .020" holes for the center bent ends of the spokes. The top layer of .032" plywood had a hole cut out to index the rim on center with two screws holding the rim flat until the spokes were soldered in place. Spoke ends were then snipped off and filed smooth to the rim. These principles can be used for any design or size of wire wheel. Keep in mind brass can be nickel or chrome plated if the final model requires this.

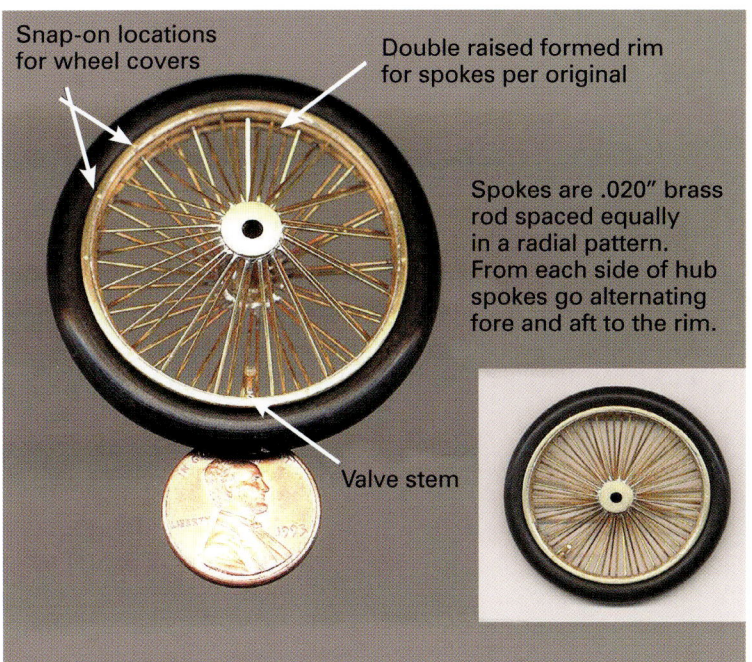

Snap-on locations for wheel covers

Double raised formed rim for spokes per original

Spokes are .020" brass rod spaced equally in a radial pattern. From each side of hub spokes go alternating fore and aft to the rim.

Valve stem

Sopwith Camel Wheel - 64 Spokes.

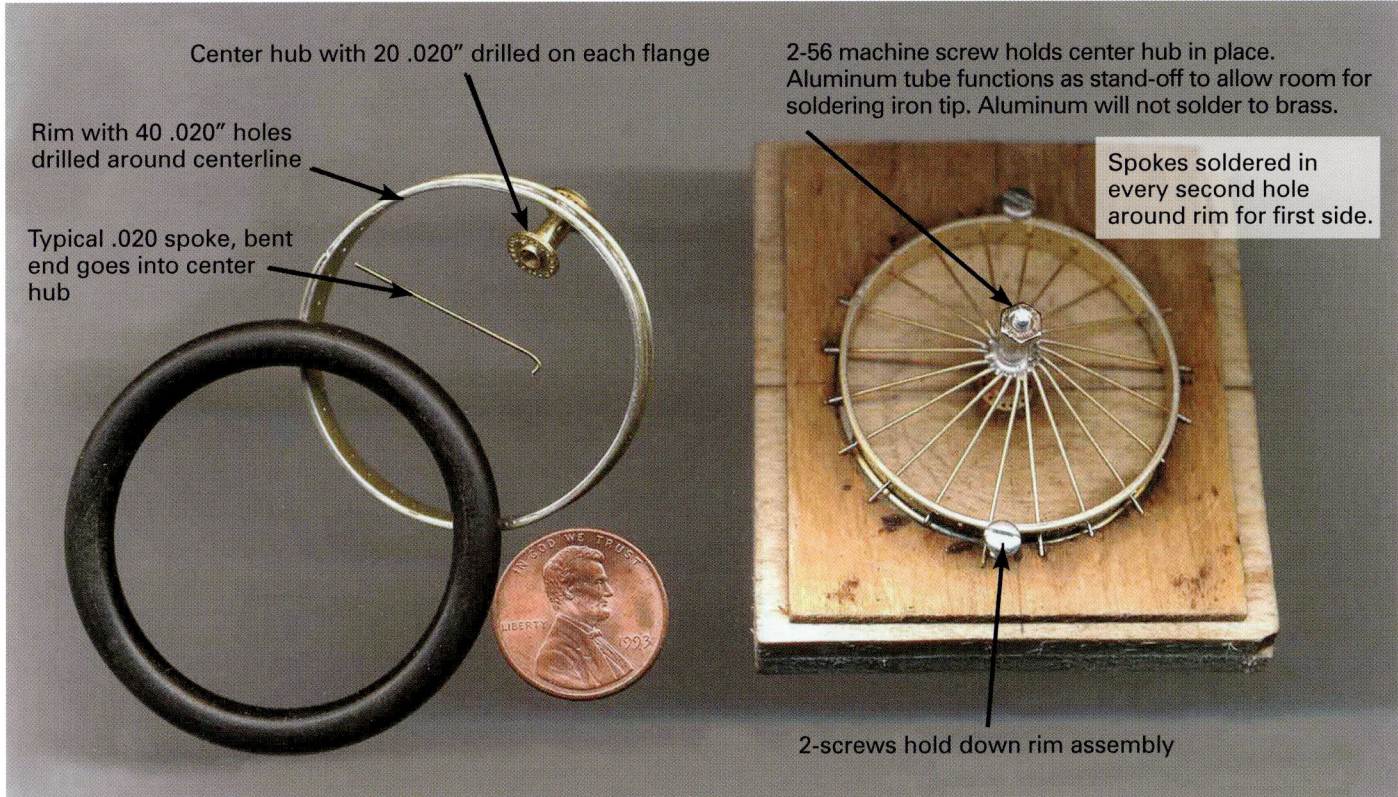

Center hub with 20 .020" drilled on each flange

2-56 machine screw holds center hub in place. Aluminum tube functions as stand-off to allow room for soldering iron tip. Aluminum will not solder to brass.

Rim with 40 .020" holes drilled around centerline

Spokes soldered in every second hole around rim for first side.

Typical .020 spoke, bent end goes into center hub

2-screws hold down rim assembly

O-ring can resist temperature of soldering iron when used as form for rim. .005" band is butt soldered inside o-ring and then a 1/32" ring is soldered to each side of the band while in wood fixture the thickness and diameter of o-ring outside diameter. On backside of this fixture.

Working with Wire

Making a wicker seat of brass wire is quite feasible, while trying to maintain the correct authentic detail at 1/16th scale. Using .012" brass wire, every detail of the original seat was able to be duplicated. The seat frame was formed using .062" round tube. The risers were formed using .020" rod inserted into holes drilled in the frame and soldered together. Brass wire, .012" in thickness, was cut to 20" lengths. These were loosely wrapped into .50" bundles that were then carefully annealed before weaving through the frame. A low flame torch was used for the annealing.

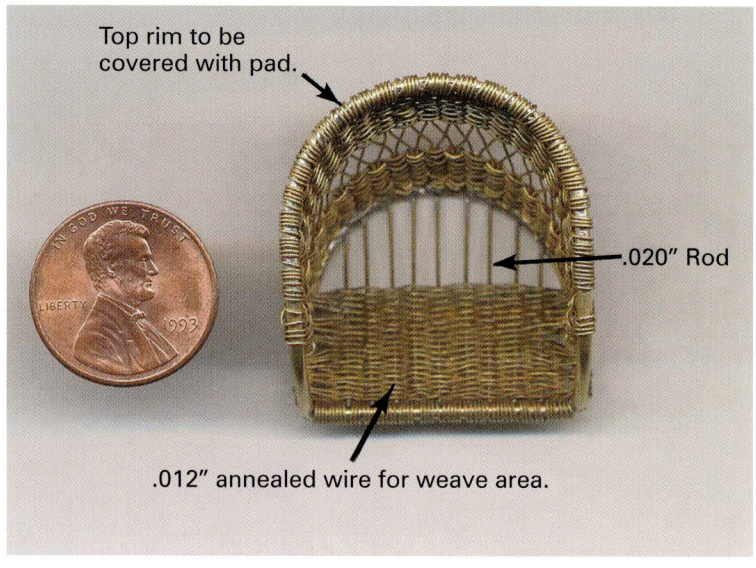

Frame formed of 1/16" tube, drilled with twenty-three .045" holes evenly spaced to receive two .020 risers in each hole.

Risers .020 rod soldered first before starting weaving step.

Wicker weave .012 brass wire that has been annealed to make more malleable to conform to weave pattern.

To anneal wire: Wrap a length of wire around a 1/2" diameter object to form a bundle. Bind bundle with end of wire to keep together. Heat evenly with flame until entire bundle almost glows and then quench in water. Wire should now be dead soft. It is better to do several small bundles rather than one large one due to uneven heating leading to inconsistency in annealing.

Sopwith Camel Pilot Wicker Seat—In Progress.

The annealing process allows the brass wire to be very malleable, forming uniformly around the risers. The lower portion of the back was woven from the bottom up to the ventilation opening, then from the top of the ventilation opening up to the top of the frame, wrapping the ends around the frame as per the original seat. The key to accurate weave detail was to pull the wire tight without breaking the wire as it was woven.

Top rim to be covered with pad.

.020" Rod

.012" annealed wire for weave area.

Sopwith Camel Wicker Pilot Seat—In Progress.

Here is the finished woven seat with the attachment pads soldered in place on the bottom of the seat. This was used as a master pattern to cast brass seats. The weave areas were filled with beeswax and burnished with a soft cloth to assure a smooth, clean surface. Once cast, the seat was immersed in boiling water, and the beeswax melted away, restoring the seat to its original finish with gaps in the weave.

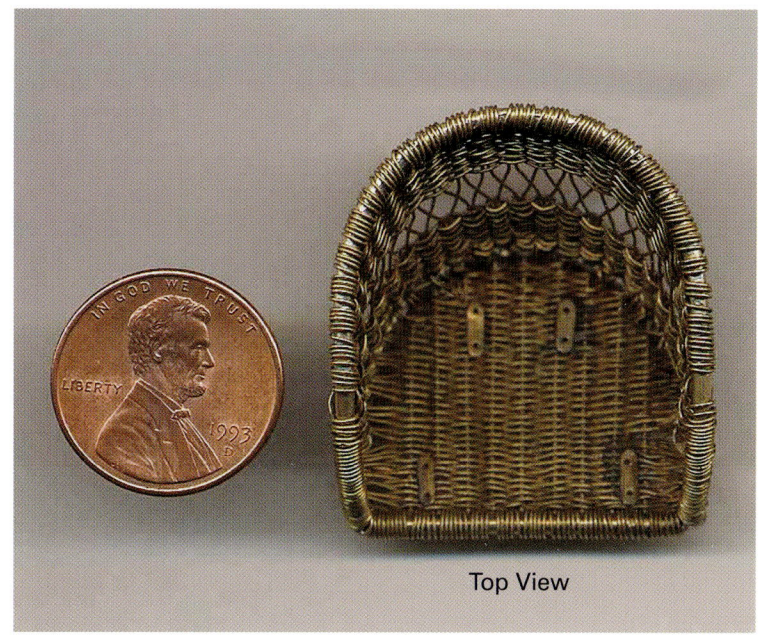

Sopwith Camel Wicker Pilot Seat—All Brass.

On the left is a cast brass seat, and the handmade master on the right. On the cast seat, the ventilation area was photo etched and soldered in place.

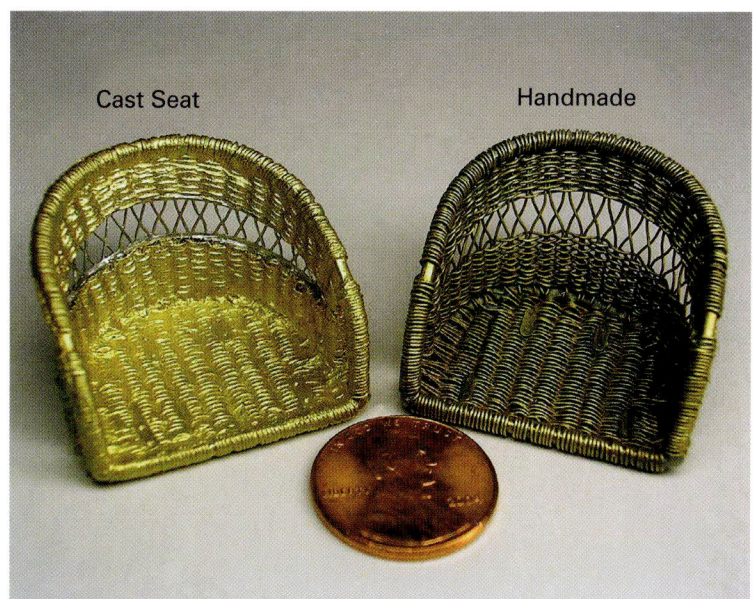

Chapter 6
Micro Hardware

Whether enhancing a plastic model kit or scratch building a model, there comes a time when nuts, bolts, and other scale hardware are needed to further enhance detail or hold parts of a model together. Best results are attained by using micro hardware, whether simulated or functional. The use of functional micro hardware requires the ability to cut both internal and external threads into parts and materials. Tools to cut the threads are called taps and dies in conjunction with cutting fluid as a lubricant.

Taps

Taps are used to cut threads on the inside of holes to thread a bolt or machine screw into or hold parts together. Taps are available in a broad range of sizes, in both course and fine threads, and in standard inch and metric increments. Taps require a very specific drill hole size to cut the threads into. Refer to a tap chart for the correct size. If a hole is drilled too small, the tap can be broken. Taps are made of hardened tool steel and are very brittle. Use cutting fluid as a lubricant when cutting the threads. The tap is tightened in the square end of the T handle, inserted perpendicular in the hole, twisted a few turns clockwise and reverse twisted to clean the chips out. Continue twisting clockwise again slowly a few threads at a time and add cutting fluid as needed until the tap penetrates the hole. The tap should turn freely with no resistance when cut properly. Work slowly with very small taps. For taps with no square end, use a pin vise instead of a T handle. In the picture, the smallest tap is size 000-120 for reference.

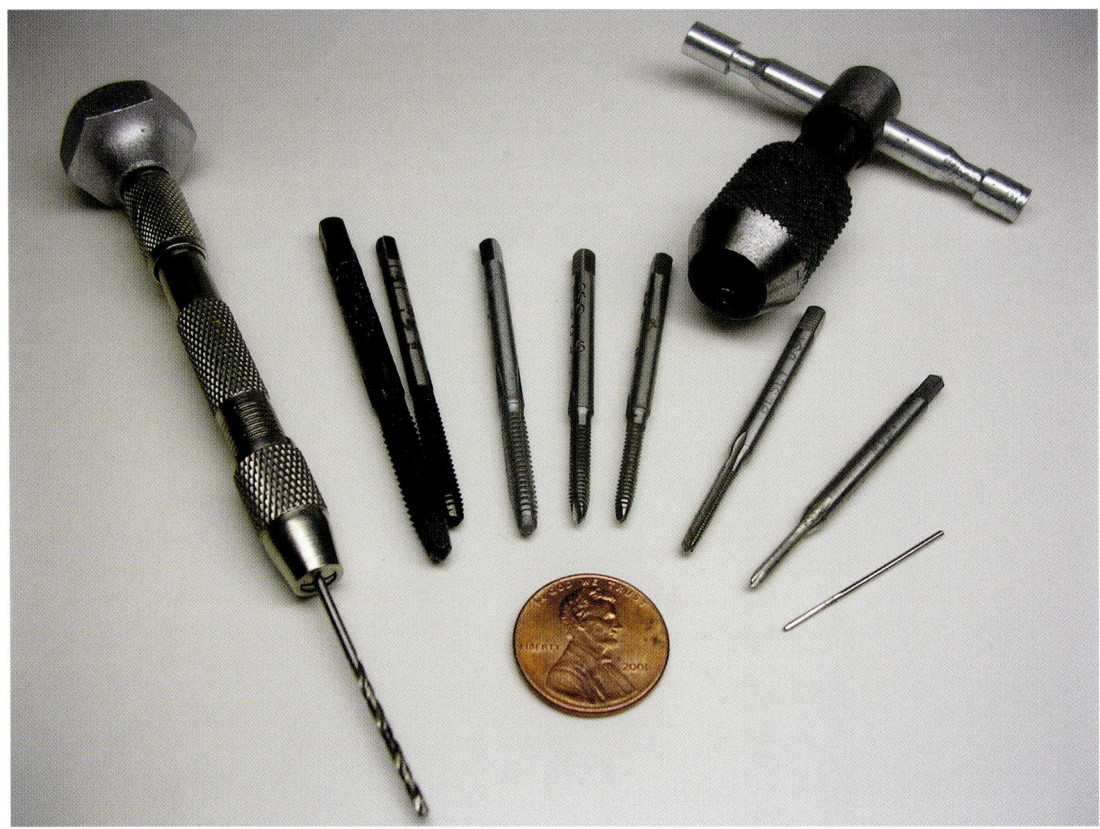

Here is an example of the benefit of tapping threads. The four large countersunk holes on top of the engine head have been tapped with 3-56 threads for the spark plugs to be threaded into.

Soldered assembly cleaned up with file and test fitted to block.
Spark plug holes threaded 3-56.

Wood can also have threads tapped into holes. Hardwoods are best for this. Once the threads are cut with the tap, add a drop of Super Glue into the hole and allow some time to set and dry. Once set, chase the cut threads with the tap again. The Super Glue hardens the wood fibers where the threads were cut. In this example, a valve stem has been screwed into a hard Maple wheel rim, and the valve stem cover can be unscrewed from the valve stem.

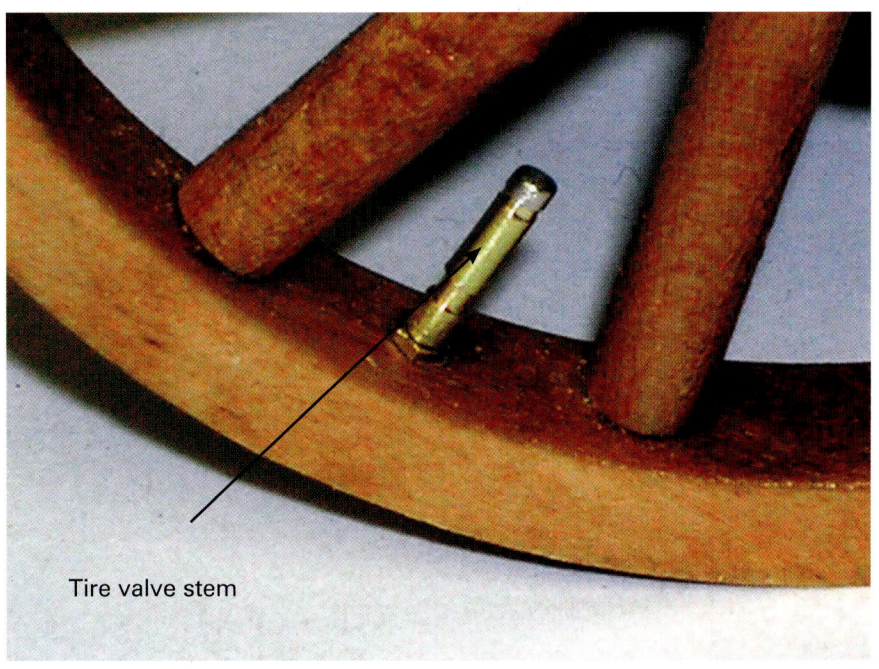

Tire valve stem

Dies

Dies are the opposite of taps and cut threads on the outside surface of a rod or tube. They are available in sizes corresponding to taps. Like taps, there are specific outside diameters of the rod or tube needed to cut the proper threads. Suppliers of taps and dies usually offer reference charts, or they can be downloaded from the internet. Dies are available in round and hexagon designs. Both may be seen in the picture below. The die indexes into the handle and is locked in place by aligning the recessed hole on the side of the die with the set screw in the handle. Note that on one side of the die, the thread cutter hole is wider than the other. The wider side is set squarely on the end of the rod, and with cutting fluid and downward pressure it is slowly and carefully twisted clockwise to start the thread cut. Like taps, turn a few twists at a time, then back off to clean the chips. The die should turn freely when completed. Some dies also have the feature to be able to adjust the thread depth and purchase by tightening a screw in the side of the die, as seen on the round die in the picture.

The threads on these leaf spring clamps were cut using a 00-90 die. Mastering the skill of tapping and cutting threads will save hours of time and effort in model building. The key for both is to use plenty of cutting fluid, go slow and straight, especially with the small sizes.

Nuts, Bolts, and Machine Screws

Micro nuts, bolts, machine screws, washers, and other fasteners of various thread sizes and lengths are commercially available in brass, nickel plated brass, and stainless steel. While you could make all of these various hardware items with a lathe, hexagon bar stock, taps and dies, commercially-available hardware saves a great deal of time in model building. Visit the following websites for a good selection that will satisfy any modeling need that may arise. Scale Hardware Inc. also offers a good selection of simulated hardware for super detailing of non-functional nuts and bolts.

www.scalehardware.com
www.specialshapes.com
www.morris01550.com

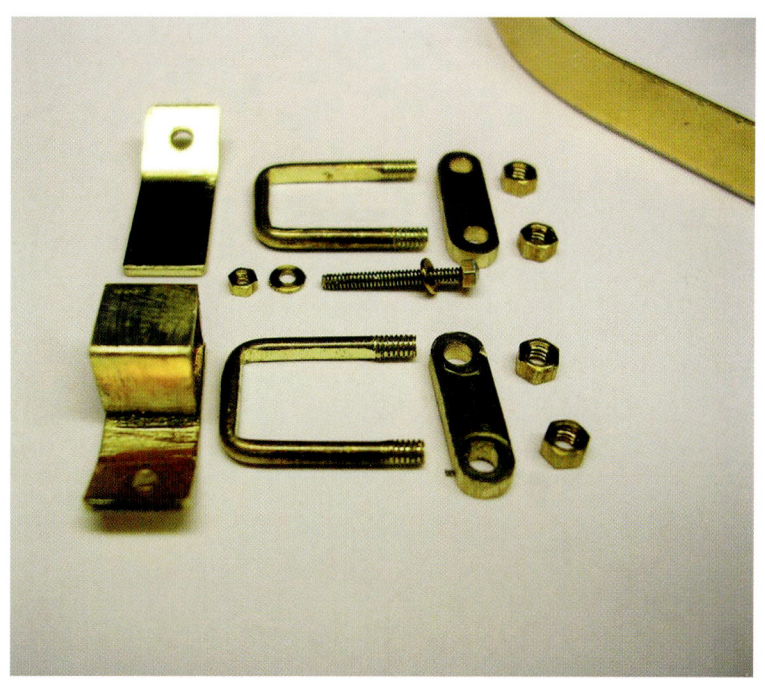

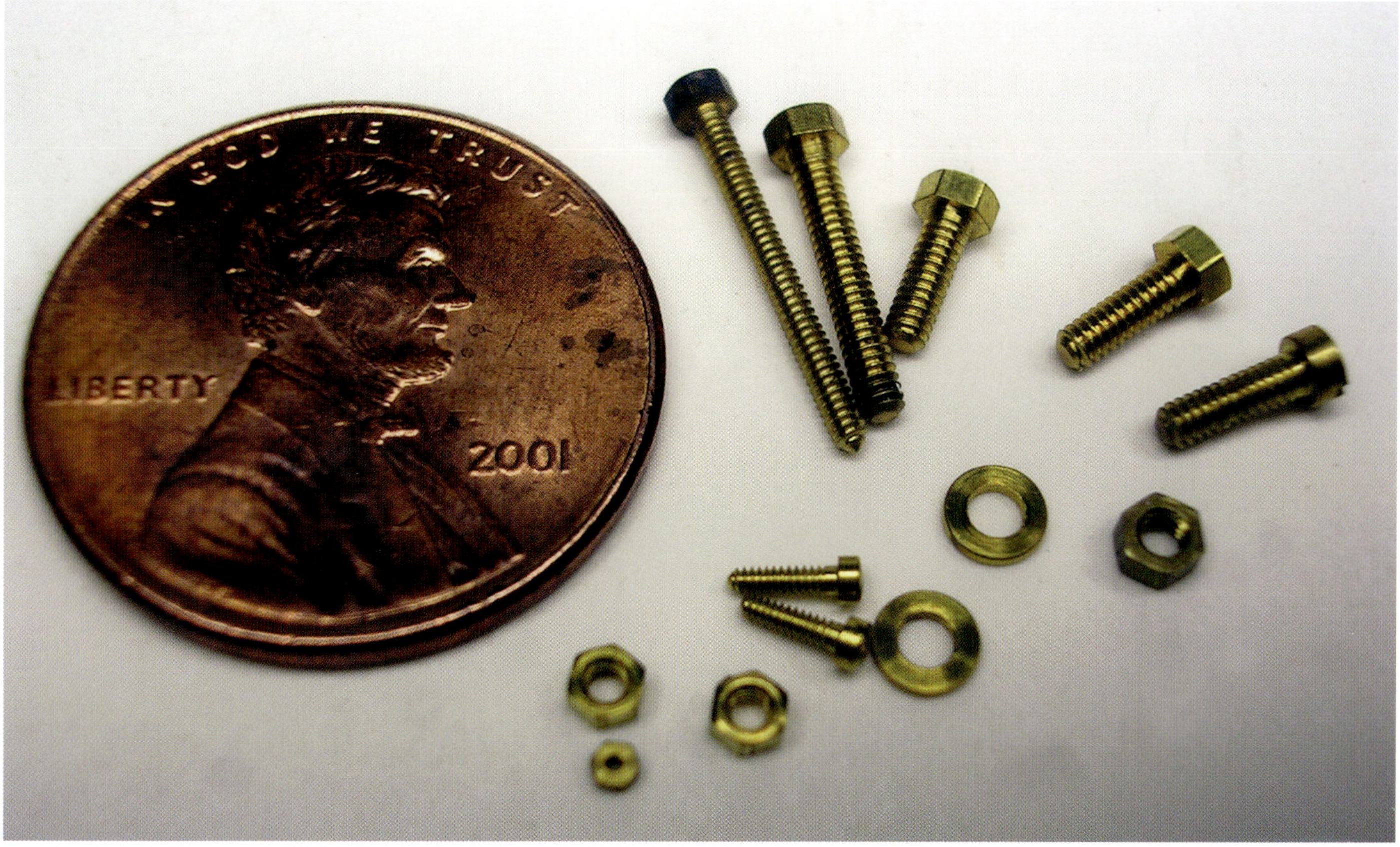

Chapter 7
Cleaning Parts

In Chapter 4, I touched on the importance of cleaning excess solder and flux from parts. This is the most important operation needed prior to painting or electroplating. The most common processes for cleaning are mechanical, chemical, and abrasive.

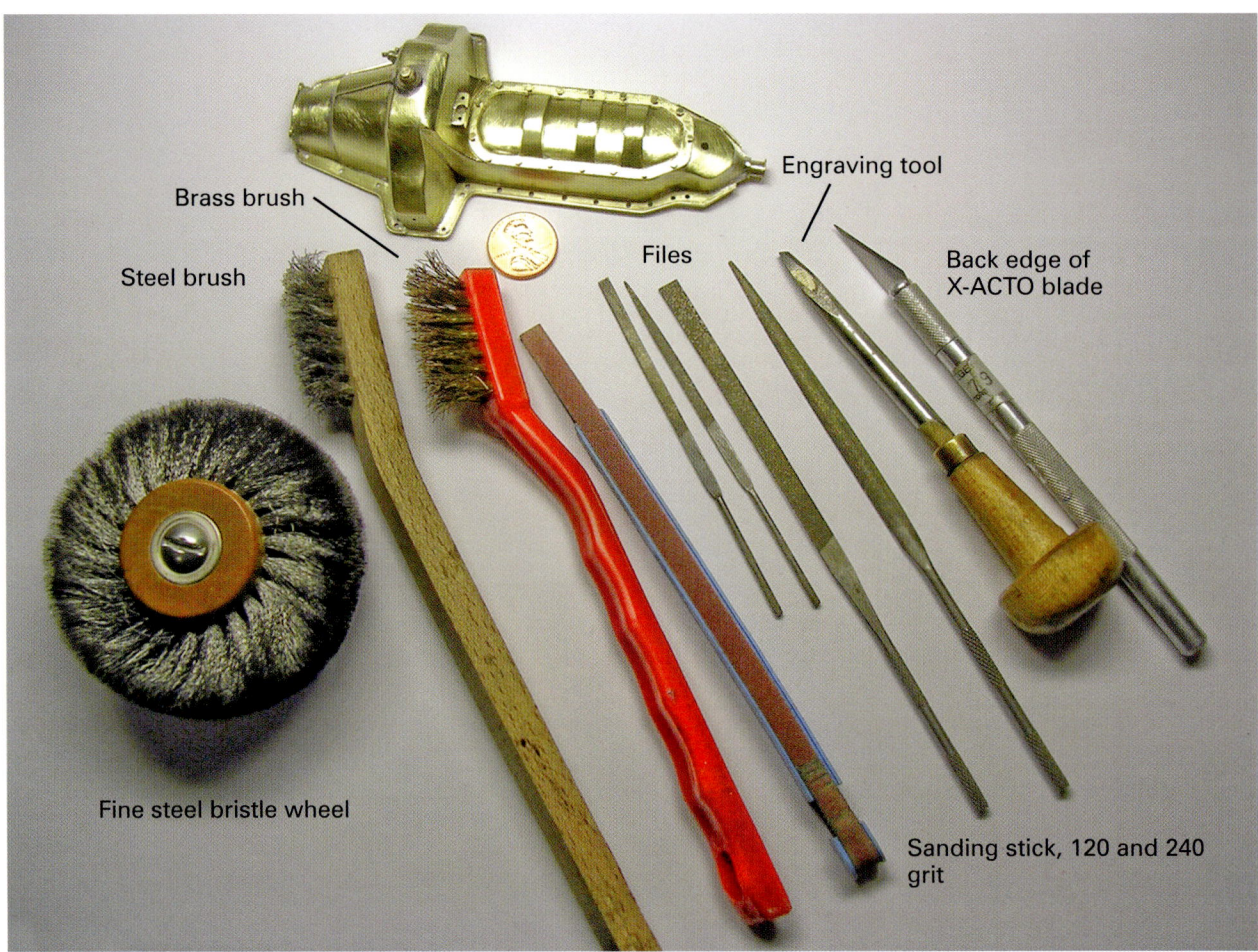

Part cleaning tools.

Mechanical cleaning is the use of tools to physically remove excess solder. The common tools used are the backside of a #11 X-ACTO® blade as a scraper moving parallel to the metal surface. Another tool is the engraving tool that is pushed while rocking back and forth or dragging. Files can also be used, varying in sizes and tooth cut. When filing solder, the coarser the teeth on the file, the easier to clean with a file card. Apply talc powder to the file teeth to reduce solder build up. After using files, sanding sticks and sandpaper of various grits work very well. Scotch-Brite® pads are available in different grits and are good for cleaning if you plan to continue to work on the part. A couple of other useful tools are both steel and brass brushes as well as rotary brushes that you can mount on a drill press or lathe to buff parts. Expect to go back and forth with these tools until satisfied with the surface finish.

There are two approaches to chemical cleaning of parts. After soldering, it is good practice to clean off any residue flux using acetone and wiping with a soaked Q-tip or dipping the part in acetone. Residue flux, if not cleaned off and neutralized, will continue to etch the brass and mar the finish. Flux will also transfer to tools and ruin them, especially needle files and saw blades.

The second method of cleaning is the removal of tarnish from brass, which is accomplished by dipping parts into JAX® Instant Brass Cleaner. The cleaner does act instantly and also imparts a chemical finish that extends the protection of the part from tarnishing. One additional benefit of the JAX® Instant Brass Cleaner is the accenting of residue solder by darkening it for ongoing removal. More than once, a part was thought to be cleaned perfectly until dipped into this cleaner.

Abrasive cleaning consists of a couple of different techniques. First is the use of an air eraser with aluminum oxide media; however, this method requires the use of an air compressor. The second technique is using a firm toothbrush and pumice with water. This method works best in a sink with a large bowl of water in the sink and slow-running water with a small dish with pumice powder in it. First, dip the part in the bowl of water as well as the toothbrush to wet the bristles of the toothbrush, then touch the ends of the bristles into the pumice powder and proceed to scrub the part as if brushing one's teeth vigorously. Then, dip the part into the water to rinse and see how the cleaning is progressing. Continue to vigorously scrub until satisfied with the part's finish. This process may also accent imperfections not seen earlier, which can then be cleaned mechanically by the appropriate method, then repeat the pumice scrubbing. Once cleaned and dried, the part can be primed and painted; electroplated or dipped in the JAX® Instant Brass Cleaner for a temporary protective coating until needed to be worked on again. Parts that are to remain a natural brass finish can be clear coated, using lacquer or urethane spray can finishes intended for this purpose.

Chapter 8
Tabletop Machining Capabilities

I will not attempt to explain the use of tabletop machines. There are books that explain the topic far better than I ever could. I consider myself to be a model maker that uses machines and who learned how to use them out of necessity. By no means do I consider myself to be a machinist. However, I would like to share the kind of parts and operations that can be made with them. In my opinion, if you do not have any of these machines, consider this to be the order in which to acquire them if you intend to become a serious modeler with brass. First is the drill press, second the lathe, and third the milling machine. Of course, there are a myriad of accessories that go with each of these. Let common sense and need dictate the acquisition of tabletop machines as you develop your comfort level and skills. When a lathe is acquired, also obtain a bench top grinder to sharpen and shape cutting tools. Other nice-to-have tabletop machines for occasional use are a combination belt/disc sander and a band saw or jig saw with an assortment of metal and wood blades.

Drill Press

The drill press is the first and most important tabletop machine to acquire. The most important features to look for in a drill press are variable speed settings and a zero capacity drill chuck. Many drill press chucks only adjust down to .046" capacities, which are not acceptable for the 61 to 80 sizes of drill bits that are often used in modeling. A pin vise can be used for the smaller drill bit sizes by being chucked in a drill press, but this reduces the working depth of the drill press. When shopping for a drill press, the chuck capacity should read 0-.025", or 0-.375". Here are random examples of parts where the drill press was used to drill location holes for pins, through holes for additional parts or pilot holes for sawing.

The drill press can be used with all other modeling materials as well.

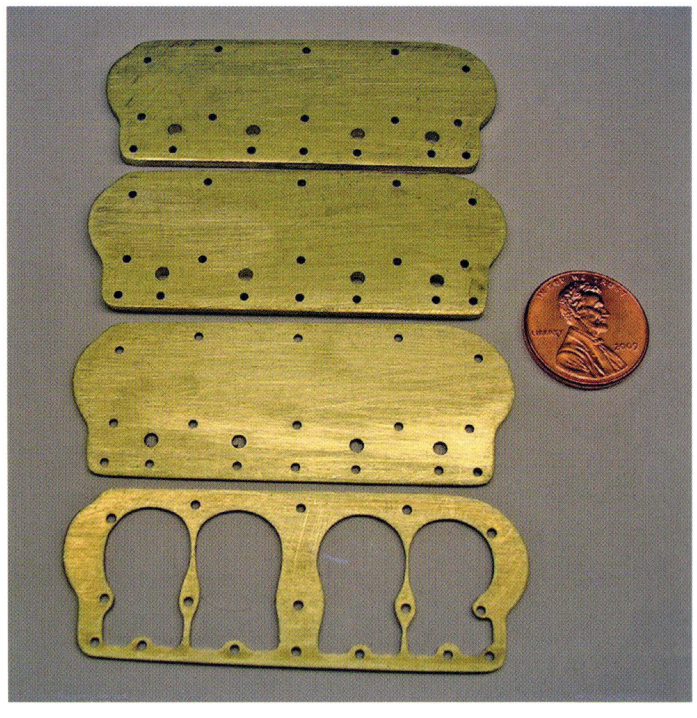

All bolt holes located on each layer. The top three layers have spark plug locations drilled.

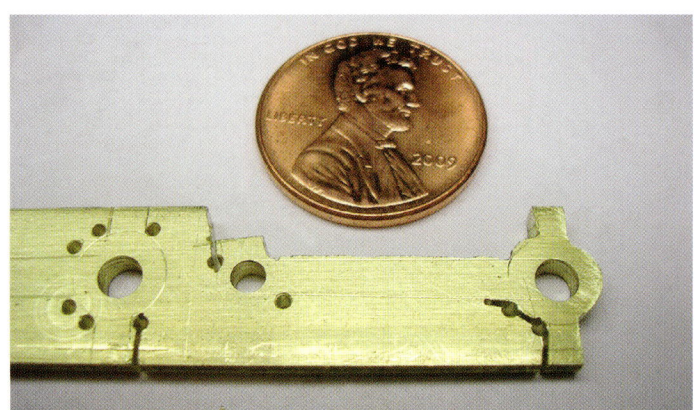

114

Tabletop Lathe

There are several manufacturers of tabletop lathes. The key points to look for when purchasing a lathe is the capacity of work that can be done, which is usually driven by the scale of models to be built. The smallest lathe I would recommend for a tabletop would be a 7" x 10". However, I consider lathes to be like garages; I have seen many garages that were too small but never one too big. Let the scale of parts to be made, accessories available, and cost drive the decision-making process. One additional thing to keep in mind about a metal lathe is that it can also be used to turn wood with the purchase of a steady rest accessory. A wood lathe cannot turn metal. Here are random examples of parts that were made using a lathe, both in metal and wood.

Crankcase turning—in progress.
Left front view.

115

Center shoulder to grip in jaw to cut off second rim

Cored out to the depth for two rims.

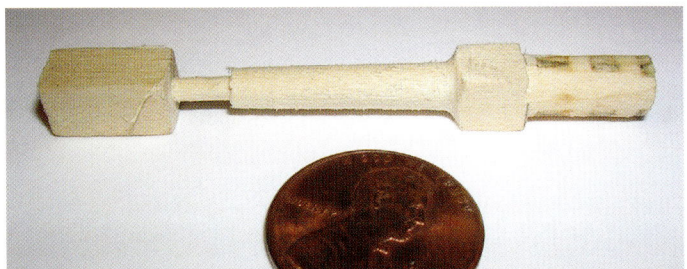

Right side view - US penny for scale.

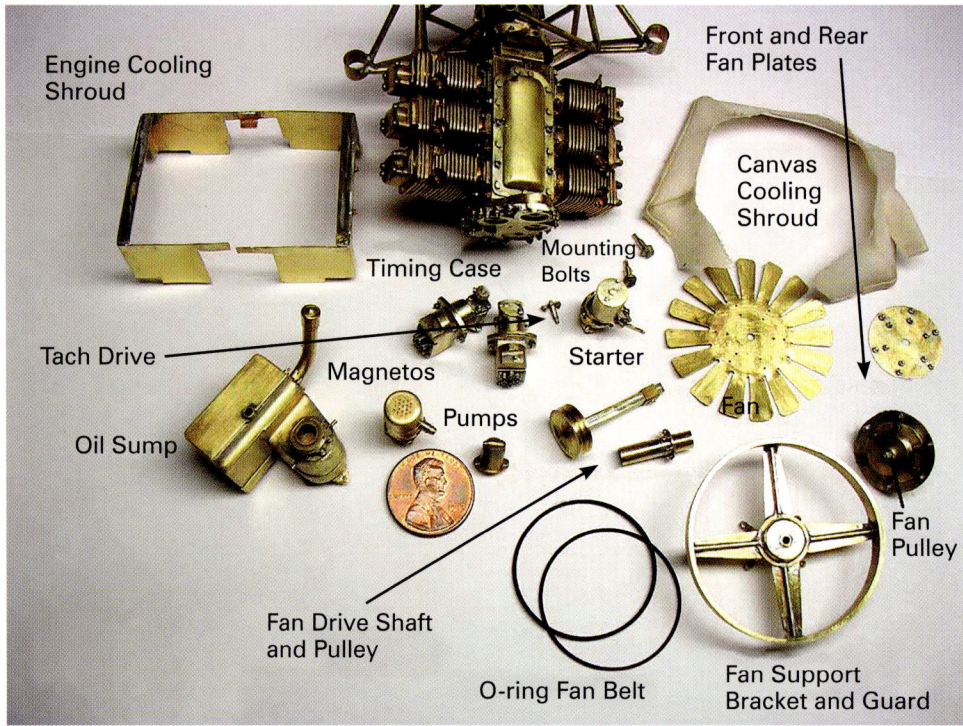

H-13D Engine Components.

Tabletop Milling Machine

Many manufacturers that offer lathes also offer milling machines. Milling machines are the most complicated and expensive to acquire along with the tools and needed accessories. It can also be the most dangerous in the hands of a novice. While learning, my best advice is to do your homework and research the operation of milling machines prior to purchase. The features of milling machines are the working capacities of all three axis of the machine. The machine size decision should be dictated by the size of the parts intended to be made. Here are random examples of parts made using a milling machine. Keep in mind that sometimes parts will require operations first done on a drill press or lathe, and then finished on a milling machine.

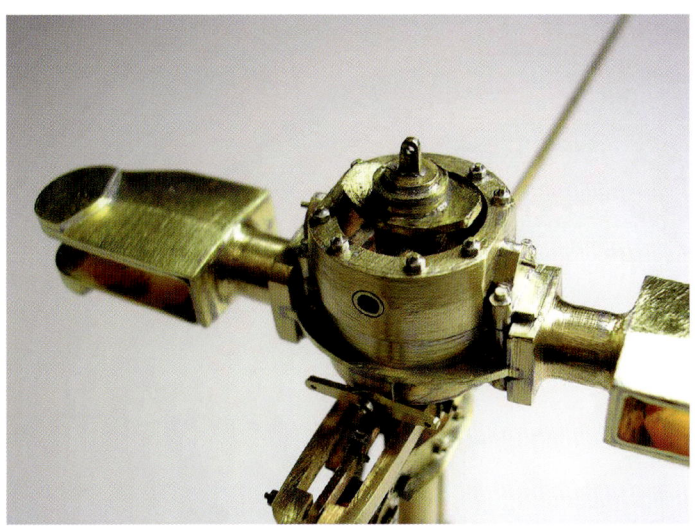

Milling pushrod channels through the cooling fins.

Side View.

Chapter 9
Working with Chemicals

Cleaning

The best chemical cleaner for brass I have used to date is JAX® Instant Brass Cleaner. Parts can be dipped into the chemical, removed immediately, rinsed in water, and they are clean and have a coating that retards tarnishing.

The setup I use is a plastic shoebox with a living hinge lid that snaps closed. The chemical has a long life and does darken over time with use. When cleaning and handling parts, wear latex gloves to dip in the cleaner and rinse in water.

Here, you can see the results of dipping the tarnished strip into the JAX® cleaner. The portion on the left is cleaned while the section on the right is still tarnished. As mentioned earlier, this cleaner will darken solders, which can be beneficial when cleaning off excess solder to clean up parts.

Electroplating

Electroplating is a chemical process in conjunction with electricity to impart one kind of metal finish onto another metal. There are commercial suppliers like Caswell Inc. that offer kits for home hobbyists. The process involves attaching a cathode clip to the part, and then immersing the part in a chemical with an anode and the reaction starts. The amount of time the process takes is based on the size and mass of the part being plated and which finish is being applied.

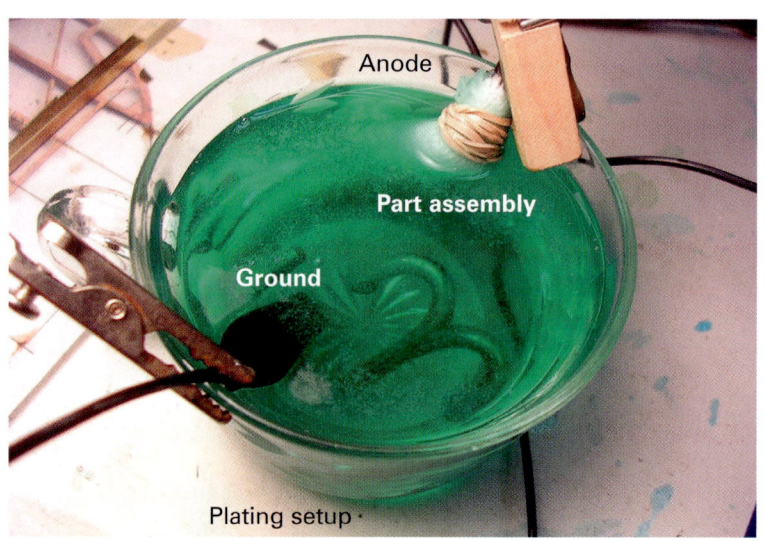
Plating setup

> **WARNING: Store chemical containers out of the reach of children and pets. Read the warning on the container label regarding the use and care of the chemical.**

The desired end result for these parts was a nickel plated finish. With brass parts, it is necessary to copper plate first to have the best results of the nickel plate finish. The copper is a better binder than raw brass. Here you can see that the complete assembly was copper plated while still bolted together.

Close-up of copper plate

Here is the nickel plated intake manifold assembly right from the nickel plating solution, rinsed in water, and not yet polished. Again, this unit was plated while assembled together.

This photo shows a fuel bowl that was copper plated by dipping the bowl in the plating solution to the top edge. It was then rinsed in water and buffed with a soft cloth. The shut off and lock nut were then plated again with the nickel solution. This multi-color plating is easily and quickly accomplished once setup.

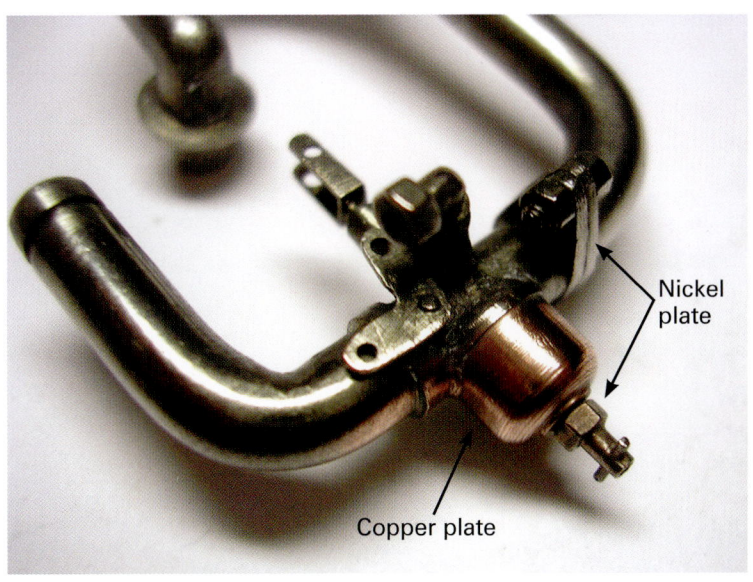

Dual plating of same part. Carburetor bowl dipped in copper plating and polished, then drain nut dipped in nickel plate.

This shows the finished, polished intake manifold test fitted in place on the engine. Notice the difference in finish between the plated manifold and the painted aluminum transmission cover.

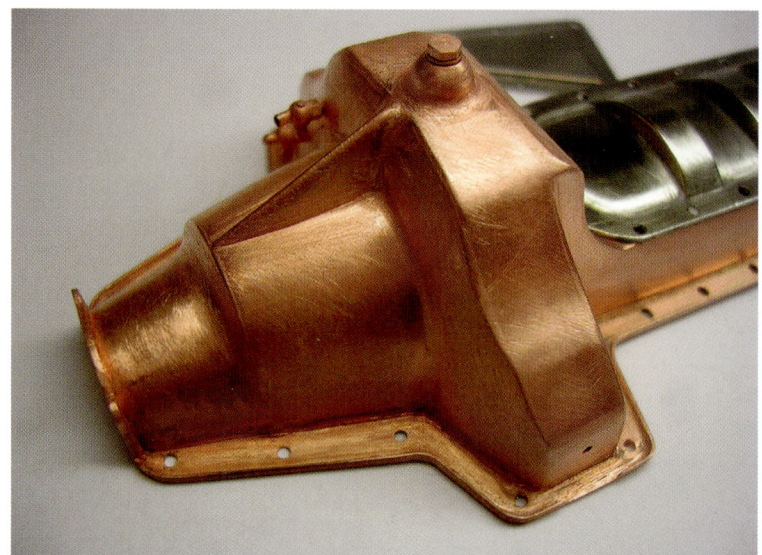

Intake manifold and carburetor were plated as an assembly—nuts, bolts, and all.

Here, the lower crankcase is seen just after being copper plated. Note how the part has a uniform copper finish covering the brass and solder joints and radii. At this stage, if a joint or part is not up to your satisfaction, it can be reworked until correct, and then plated again. The oil drain plug was plated in place. The next step, once satisfied with the part finish, would be the nickel plating of the completed crankcase.

The universal ball socket was made using a combination of brass and copper metal parts. Here you can see how nicely they finished up being nickel plated while the grease cups were left as brass parts.

Here are both halves of the rear axle unit. The brass side has been cleaned, buffed, and is ready for copper plating. The copper plated half has been carefully buffed with a soft flannel rag and just a touch of metal polish. If you are too aggressive with the polishing and buff through the copper exposing the brass, just plate again with the copper solution. There are times when a copper finish is desired, and here you can see the level of polish that can be achieved. However, once finished, the copper should be clear coated, or it will tarnish over time.

The polished nickel plated head part is resting on the upper engine block. Here, you can see the result of the copper plated, solder-filled radii where the cylinder walls blended into the engine block case.

TIP: Keep in mind you can brass plate parts to have a brass plated finish. The benefit of brass plating is that it plates the solder joints and solder fills for a uniform brass finish on the completed part.

Nickel plated head.

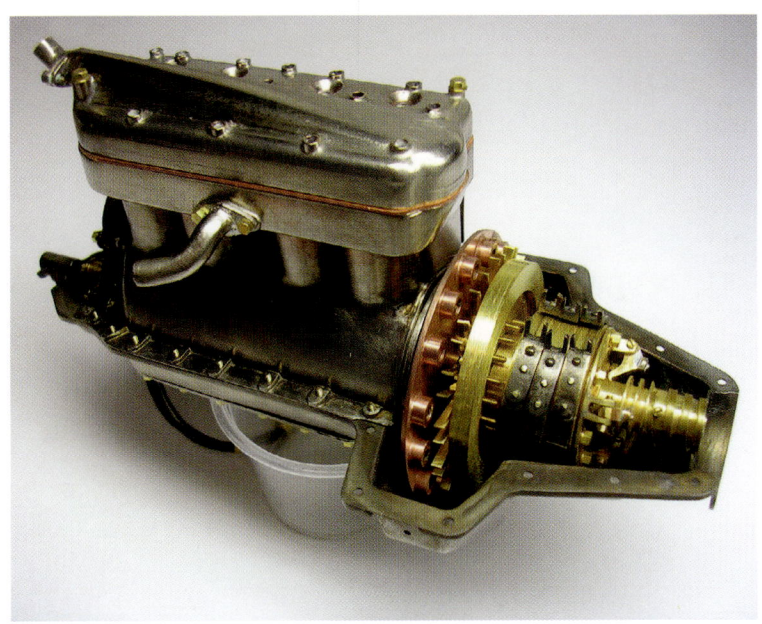

This is an example of various finished plated parts that, when combined, can have a very dramatic impact for a model. No paint can match the shine and luster of real polished metal.

The supplier of the plating kits provides clear instructions on the use of their kits. Read the instructions first. Some suppliers, like Caswell, Inc., even have online video tutorials. Below is a picture of the finished nickel plated engine. You may recognize some of the parts from previous chapters. The smallest parts that were plated in this picture are the working hose clamps from the engine to the radiator. To plate small parts like these, you will have to attach them to larger pieces of brass to provide more mass for the plating process. Otherwise, you will have darkened, burnt parts.

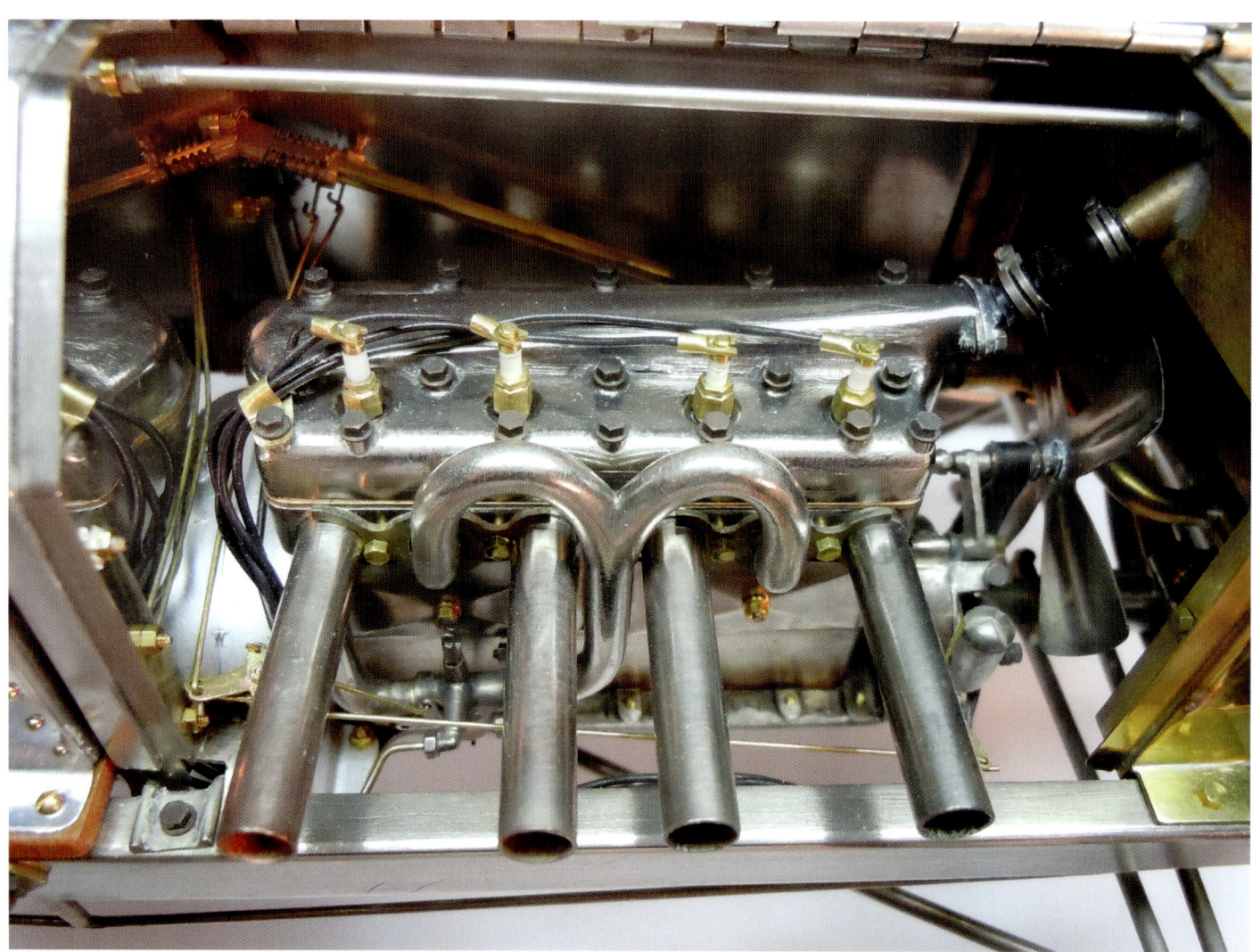

Photo Etching

Photo etching, or chemical machining, is a process for making very tiny parts that would be extremely difficult to make otherwise. There are services that can photo etch parts for you using this process, or you can also purchase photo etching kits to make your own parts. The kit instructions explain, in depth, the process, which I will not attempt to do. I just wish to expose you to, and inform you of, this valuable process for making tiny parts for detailing. Parts in the picture are examples of acid etching from both sides of the sheet. Note the little tabs that still attach the parts to the parent sheet. These are needed to hold the parts in place for easy handling once the undesired material is etched completely away.

Below is an example of one-sided photo etching to make engine name plates that shows the level of detail that can be achieved with photo etching. The original text was etched in and **MUST** is bolded for emphasis as per the original nameplate. Also, the drill hole locations for the rivets are located in the corners. The individual nameplate below is .25" wide.

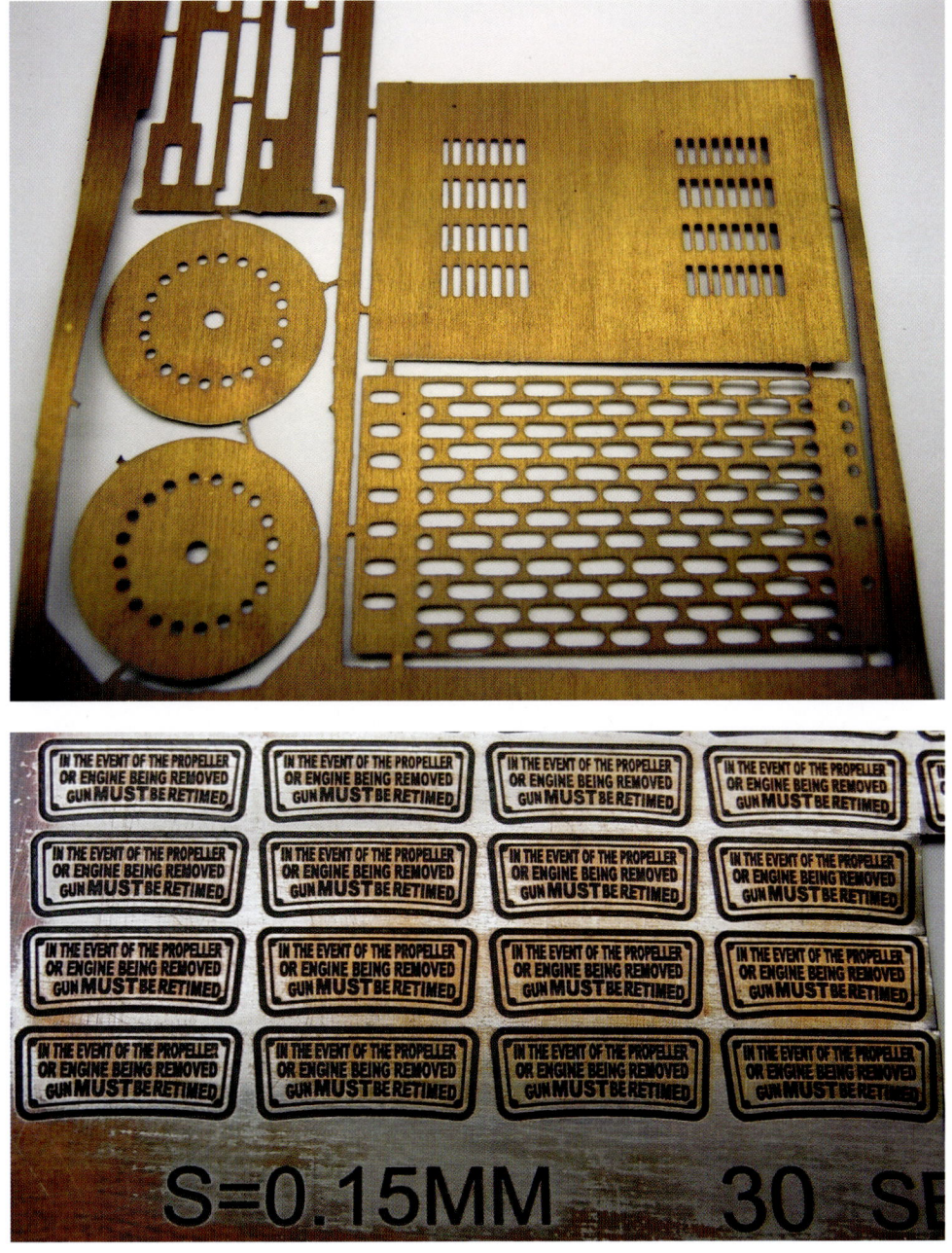

In the accompanying picture, the machine gun cooling jackets are a great example of photo etched parts that have been formed into identical finished parts.

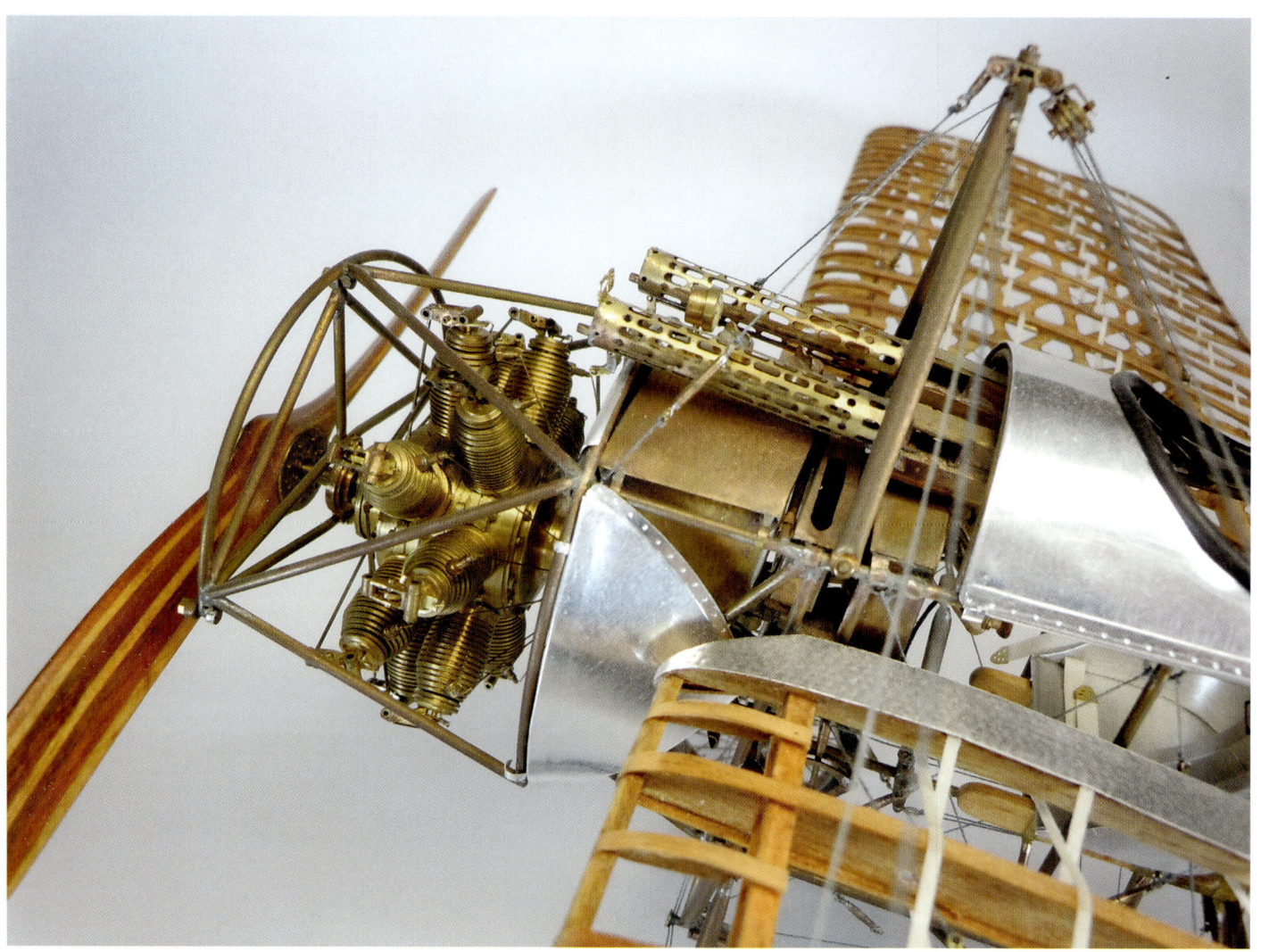

Patinas

Patinas are chemical solutions that are used to stain or tint metals. For example, in the close-up picture of the plated Model T engine, you can see the head bolts around the spark plugs are darkened using a black patina rather than paint. Brass parts can be stained with a patina using a small artist's brush or dipped in the solution and allowed to dry, then buffed lightly with a soft cloth. Repeat the process if a darker color is desired. Various patina colors are available from JAX® Chemicals, and I recommend purchasing in pint quantities—a little goes a long way. JAX® Instant Brass Cleaner also does a great job of cleaning the patina off to take parts back to their original brass finish.

Chapter 10
What to Do When You Mess Up

One of the great advantages of working with brass, besides its versatility to build with, is its forgiveness if you mess up while building a component. Sometimes, the mess up can be in the soldering process, which does take getting used to. The beauty of brass is that you can melt the solder joints without destroying the parts that were soldered together and rework them. Keep in mind unsoldering joints takes more heat. Adding flux will help in the reheating of the solder when using a soldering iron. Sometimes, it is best to use a propane torch with a low flame and the aid of picks to disassemble the parts. Parts with excess solder can be reheated and brush cleaned by using flux and an acid brush to remove the excess solder.

Another common problem is the snapping off of bolt heads by over tightening the smaller bolt sizes, usually 00-90 and smaller. The best remedy is to drill a smaller pilot hole in the end of the broken bolt stem, drill it out with the correct drill size for thread tapping, and carefully drill out the broken stem. Once cleanly drilled out, re-cut the threads using a liberal amount of cutting fluid.

Cold soldering joints are a common problem when learning to solder. The real issue is to have the patience to wait for the soldering iron to get up to temperature in the first place. A cold joint means it looks like a good solder joint but the parts are not attached together. Add more flux, and use a hot iron that is up to operating temperature for the solder being used.

Chapter 11
Bringing It All Together

The following are examples of models that I have built and their details showing the various techniques explained in *Model Building with Brass.* You will see various materials used, like wood and aluminum. A point of interest is that in building models I strive to use materials as they were used on the real subjects and incorporate as many functional features as can be made in the models. To that end, in airplanes all control surfaces work off the joystick and rudder bar. They have functional suspensions, and the rotary engines turn with the propellers. The 1911 Model T racer's hand crank turns the crankshaft, moving the pistons up and down, and, as it is geared to the camshaft, this opens the valves as well. The Model T also has a working universal joint and geared rear end that drives the rear wheels. The steering wheel functions—turning the front wheels—and has a working suspension. The hand brake operates the working rear brakes. The entire car and engine can be disassembled to all the major components and parts.

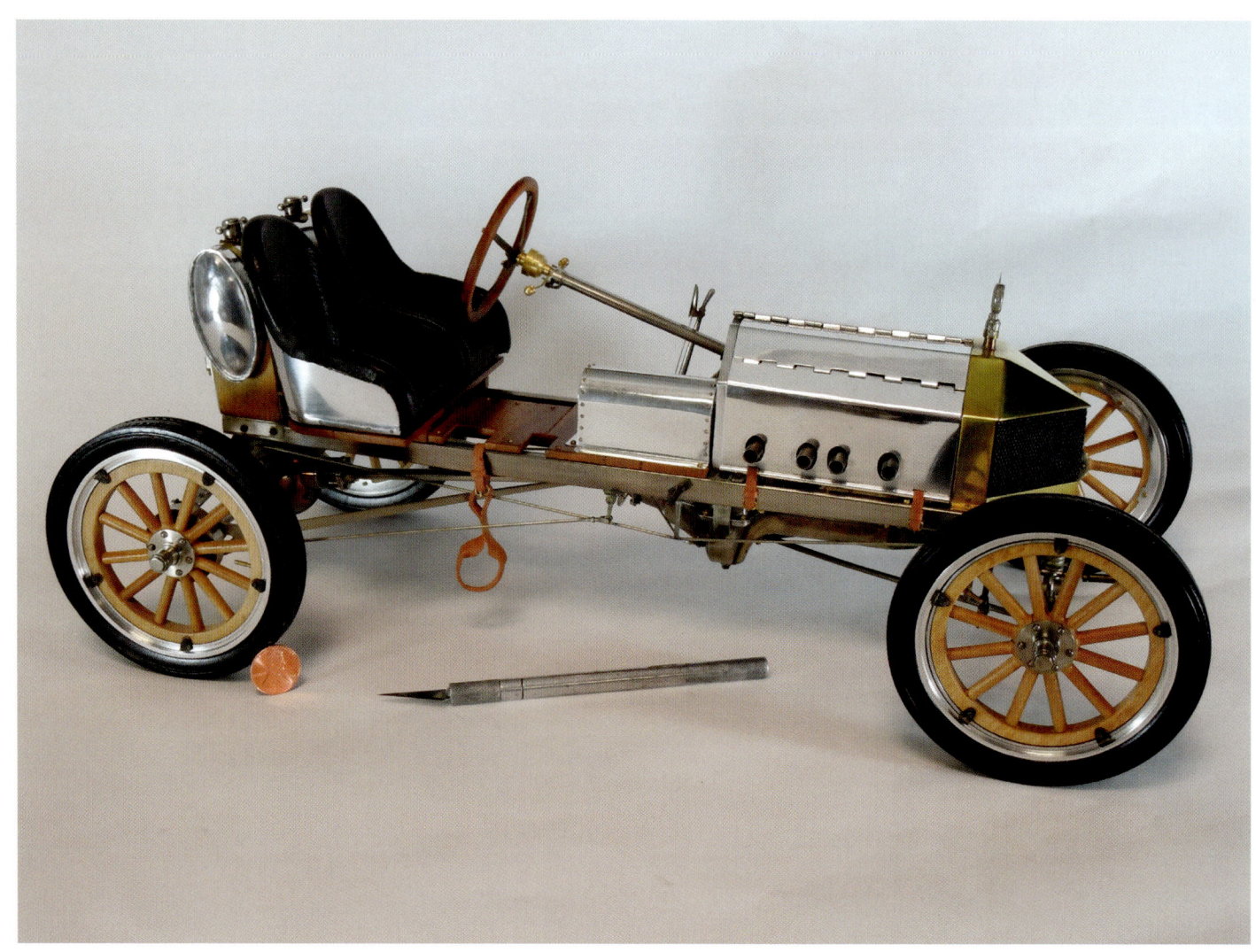

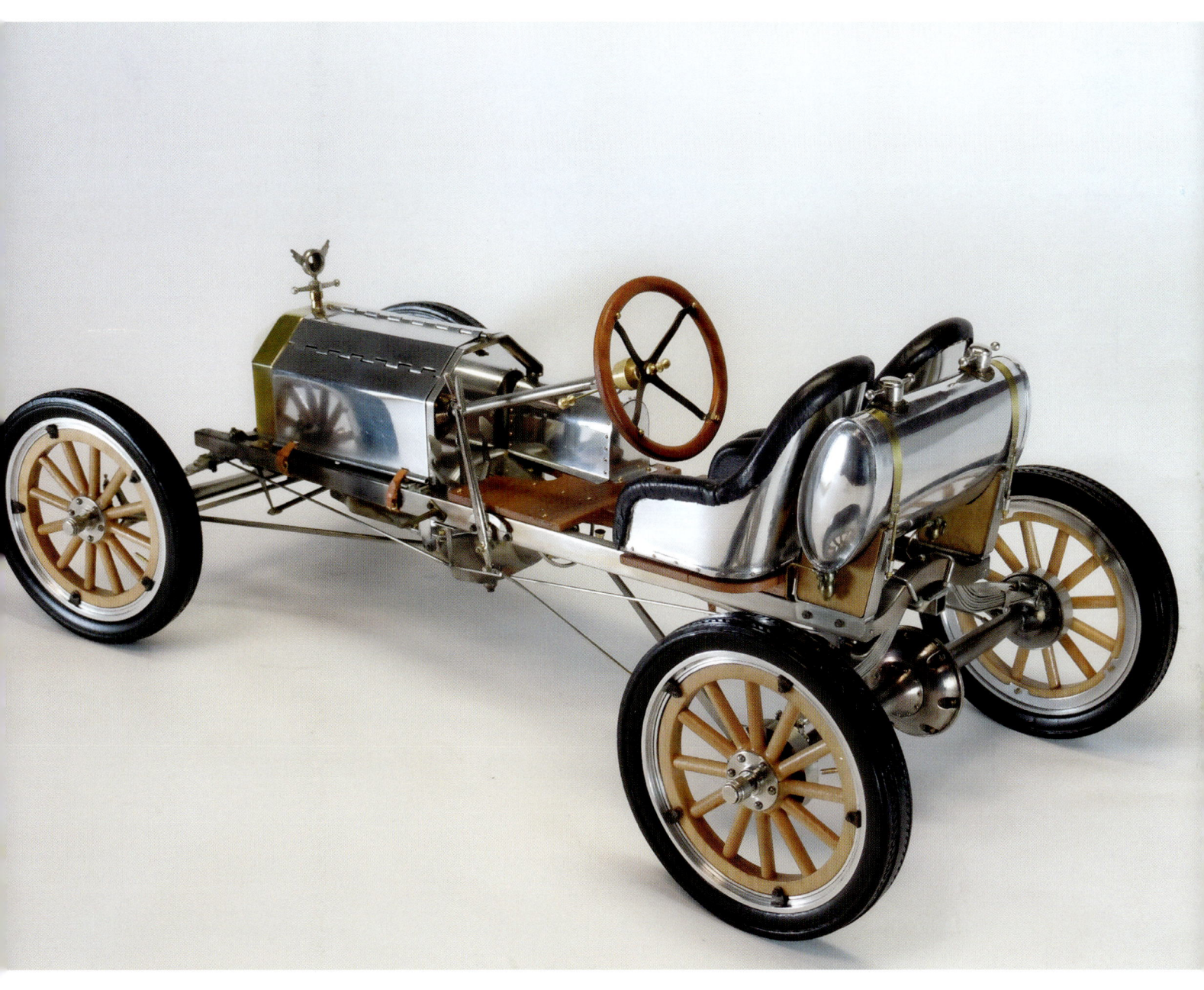

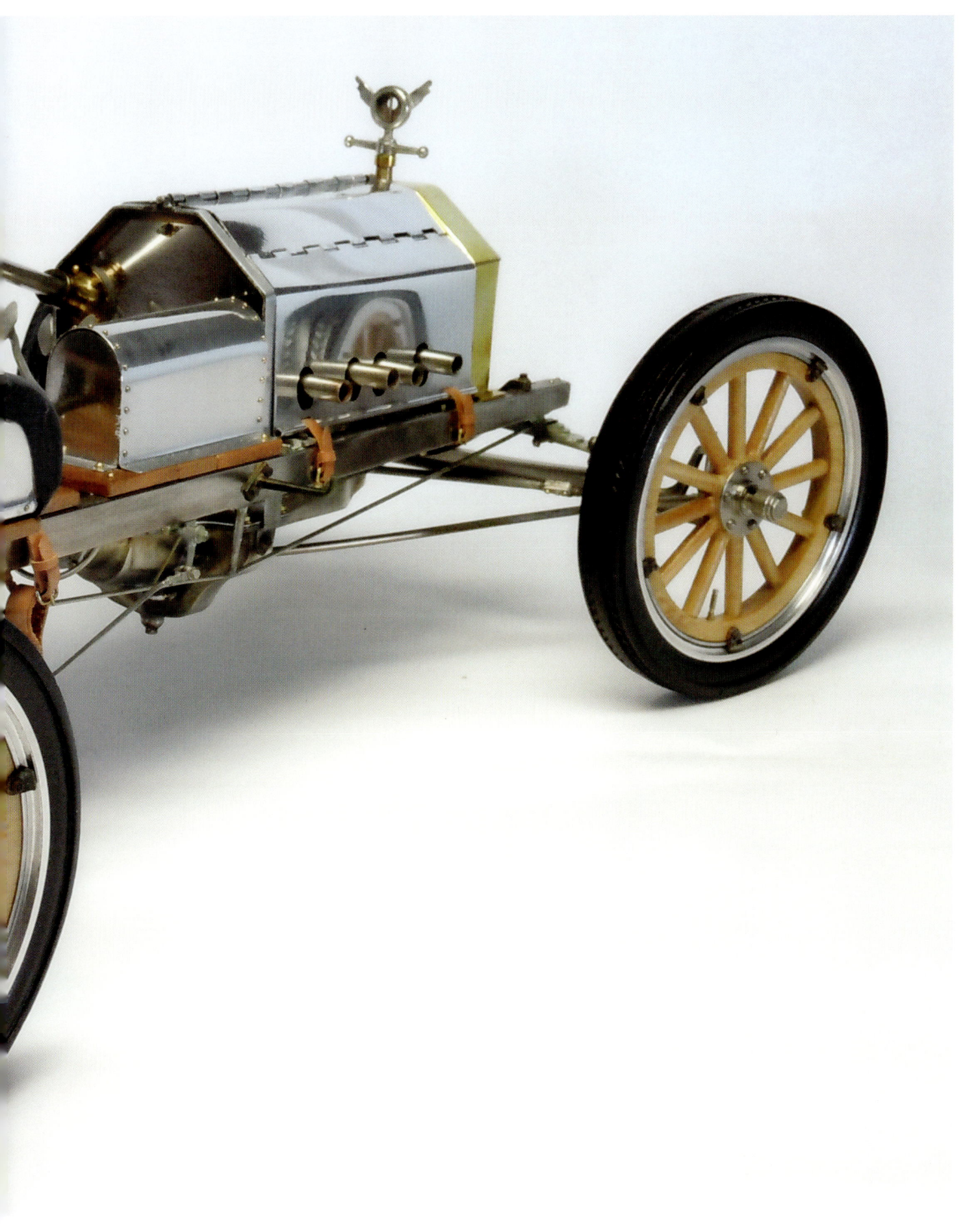

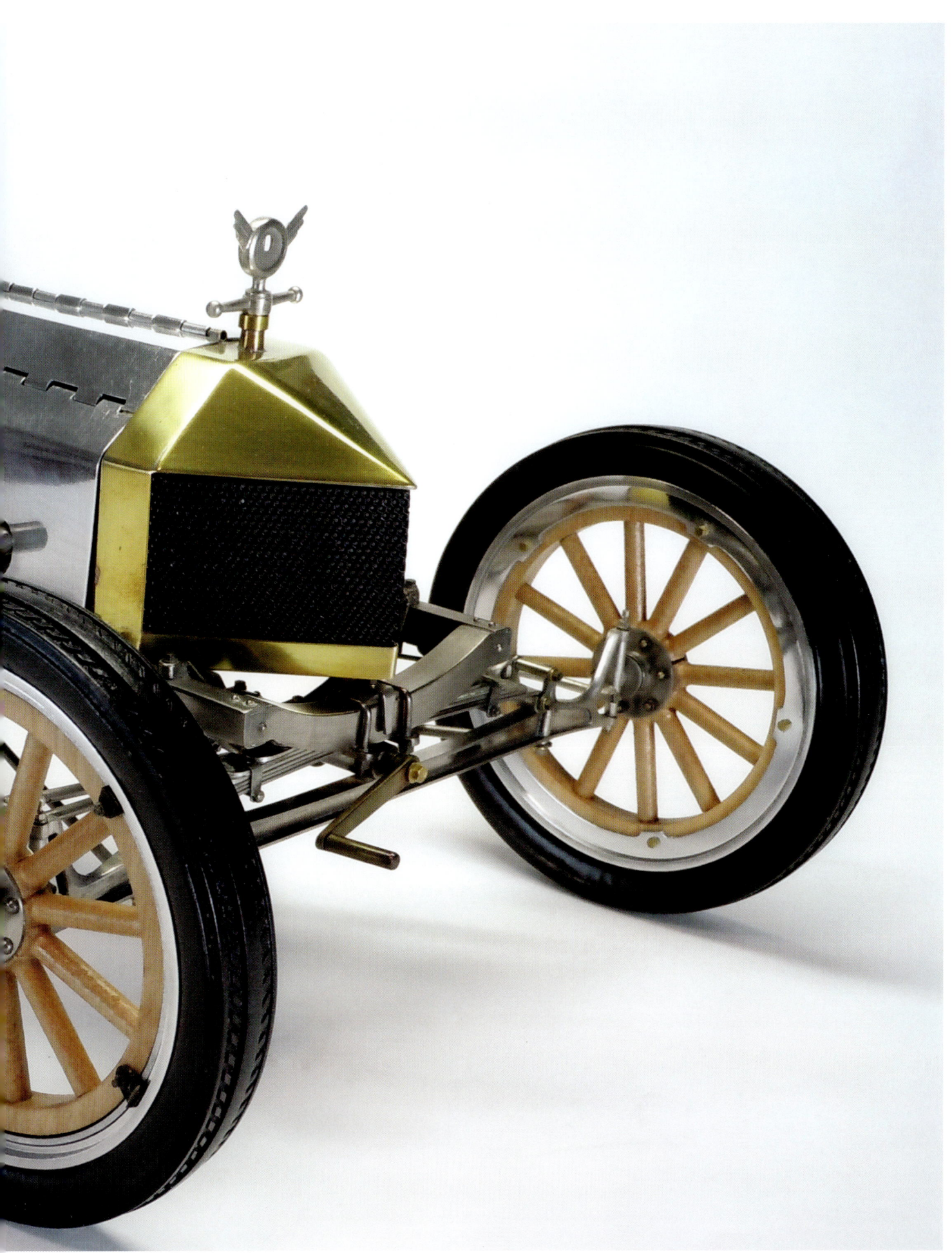

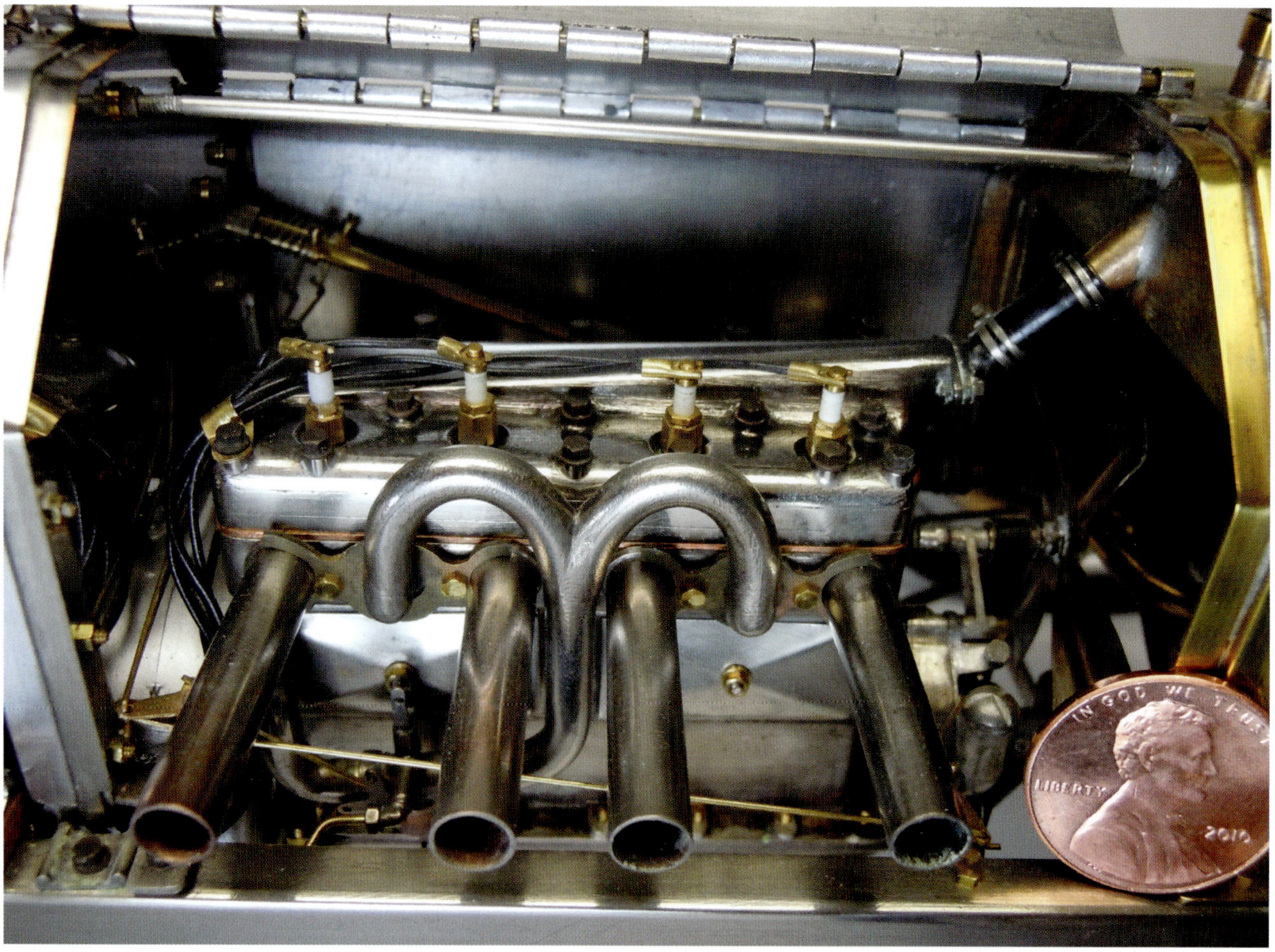
IN GOD WE TRUST
LIBERTY
2010

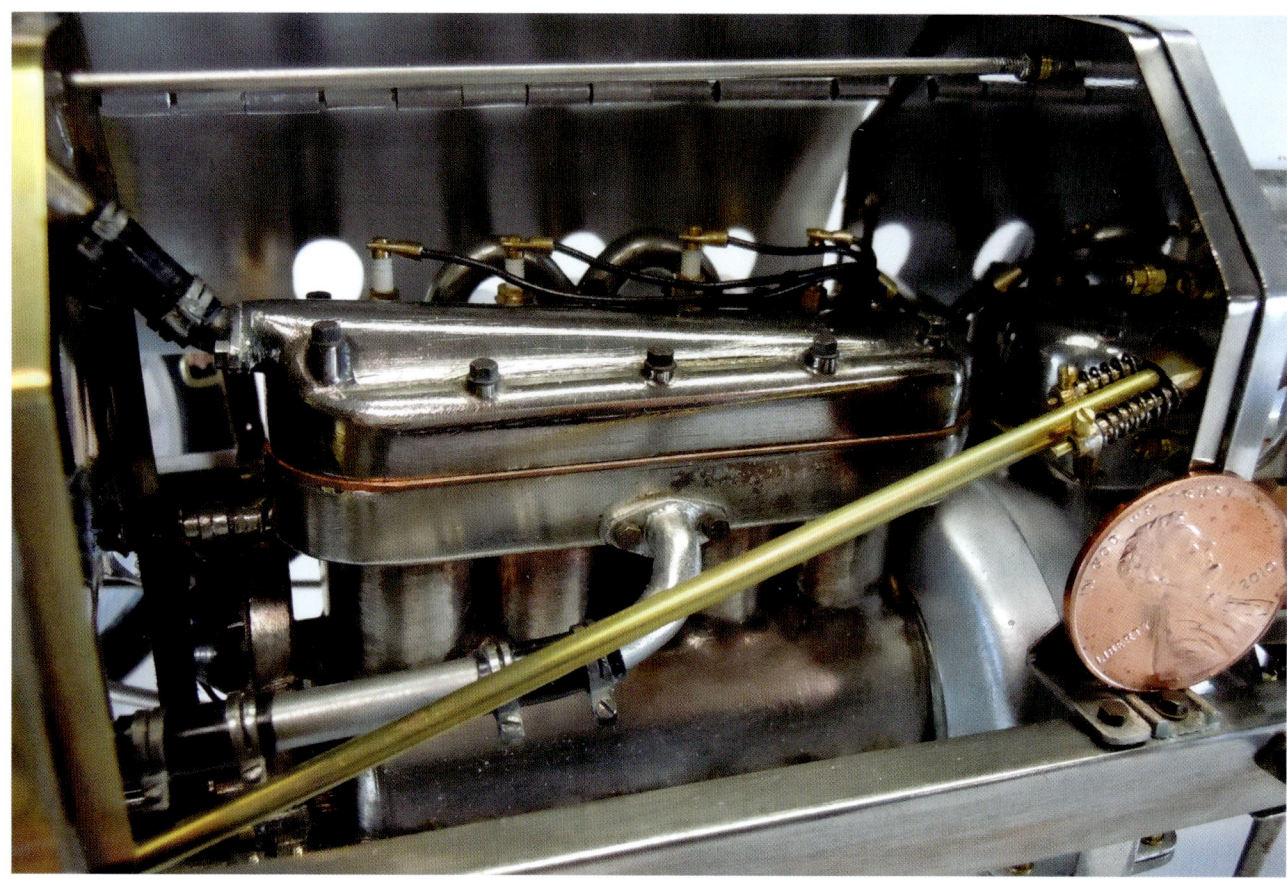

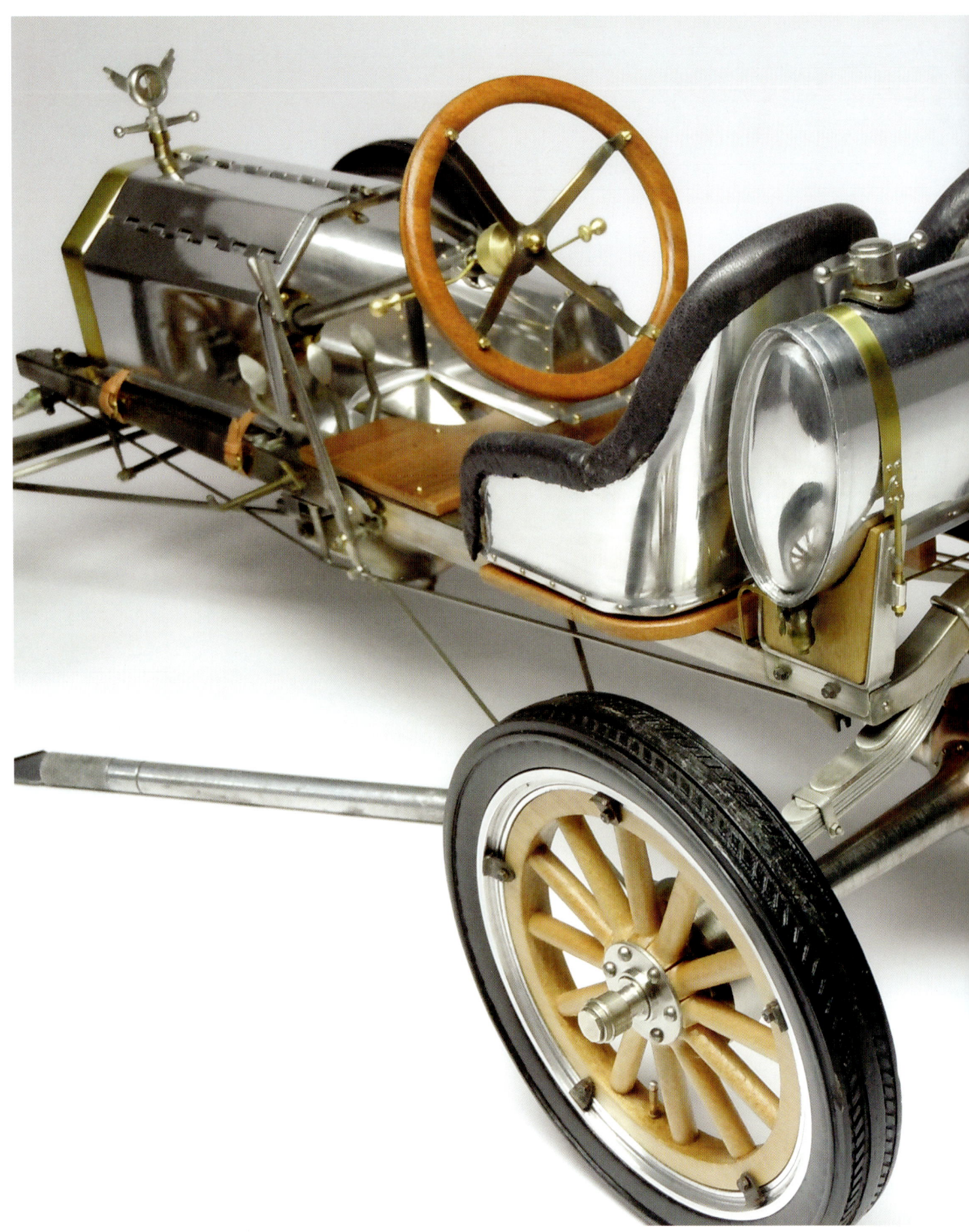

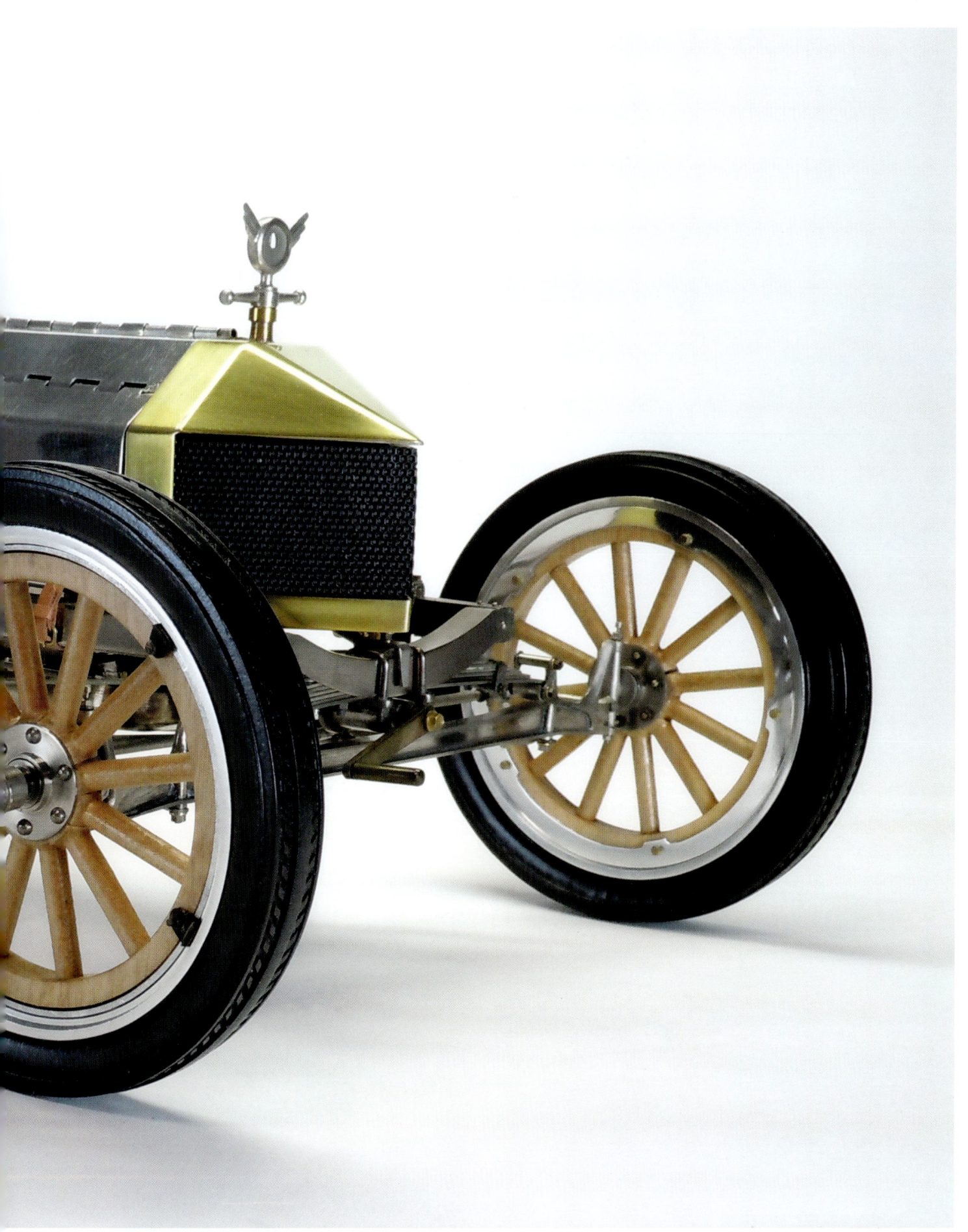

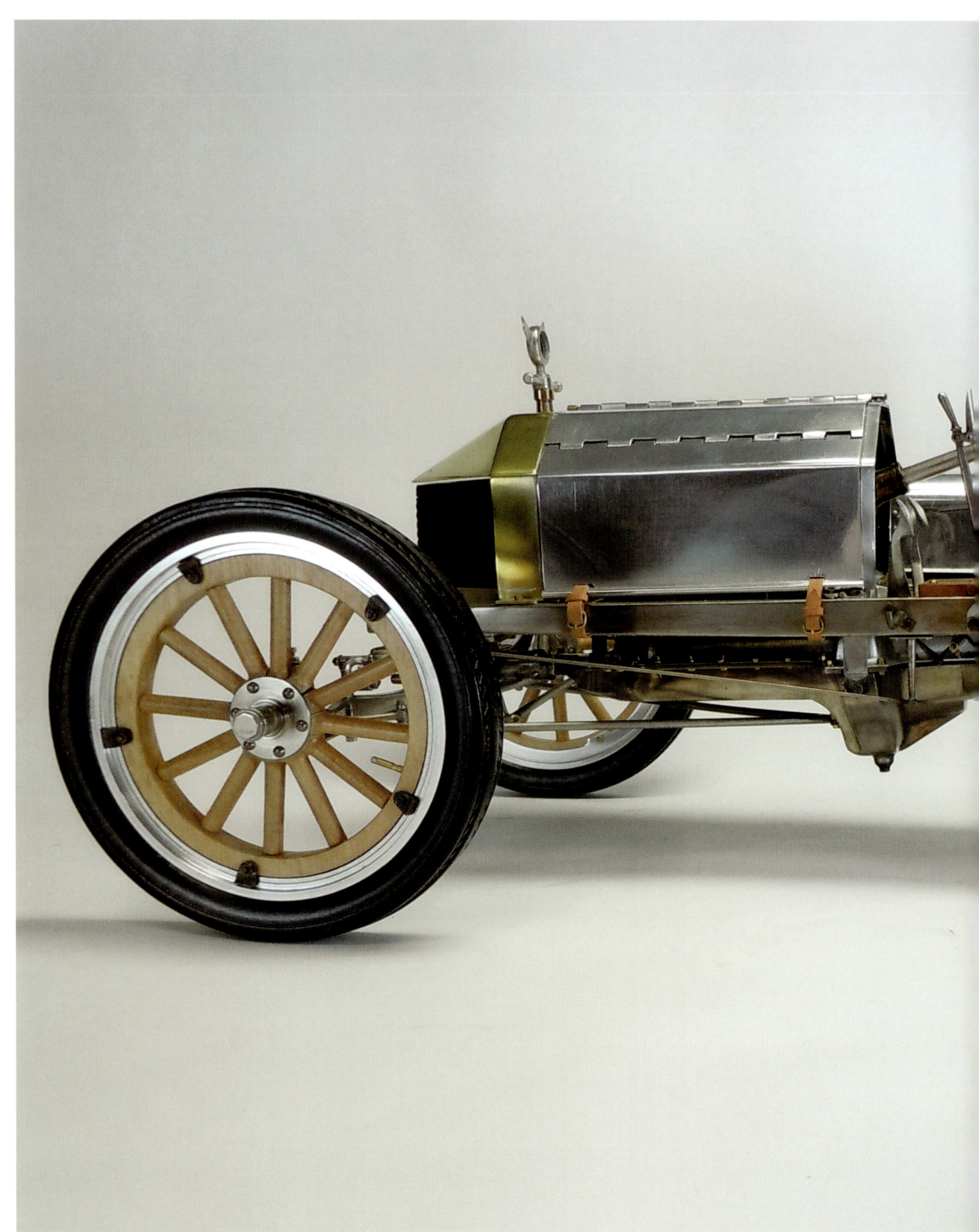

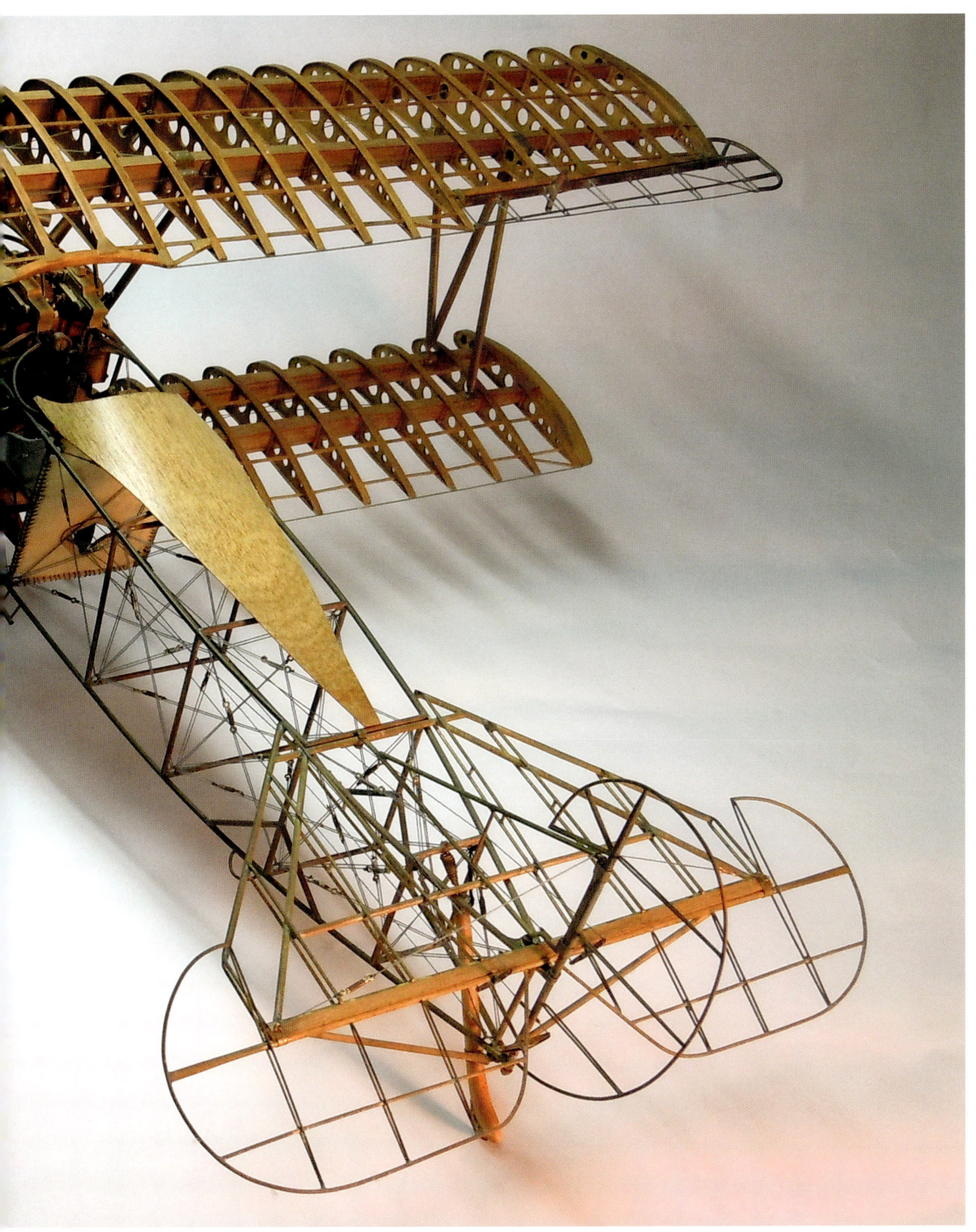

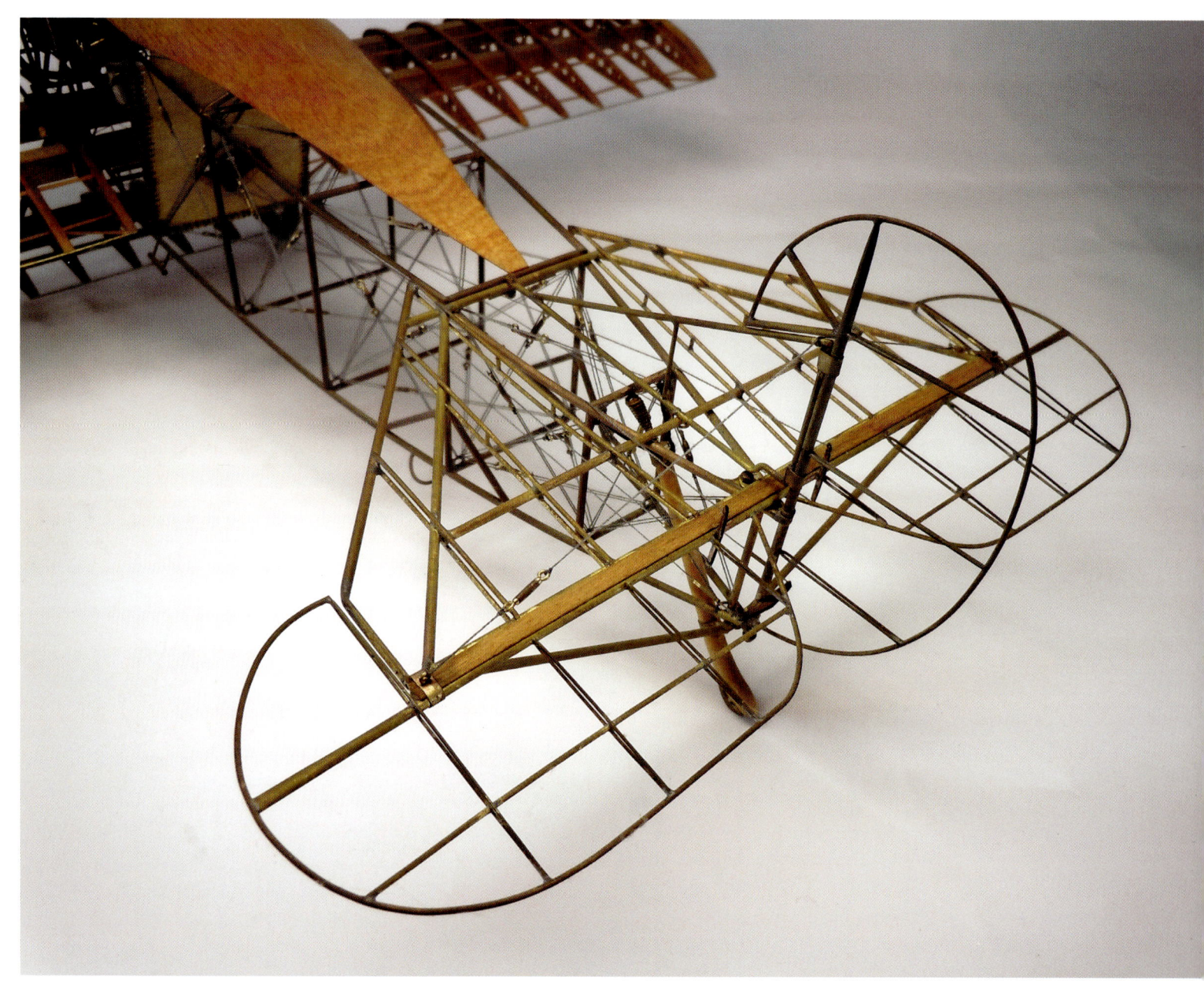

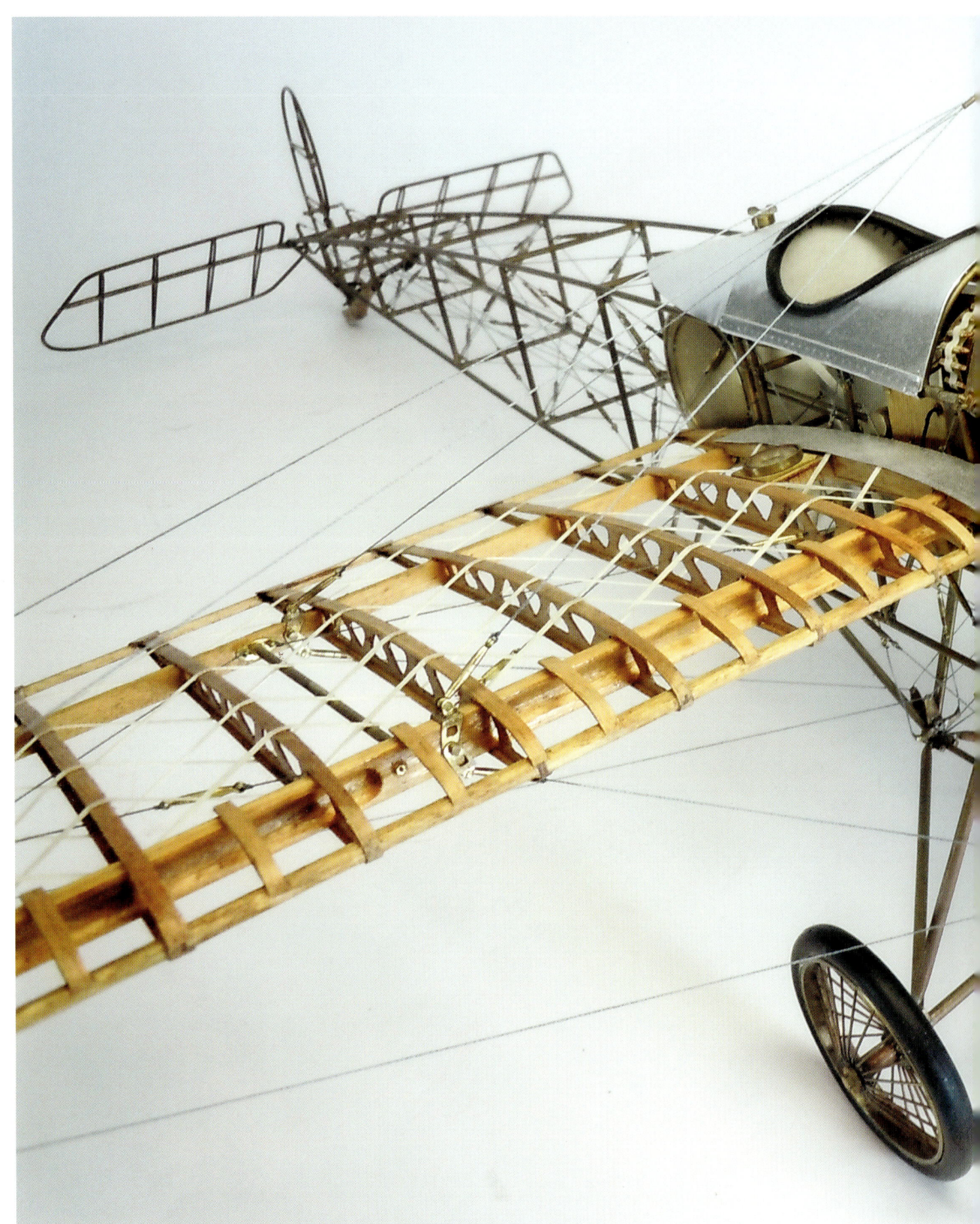

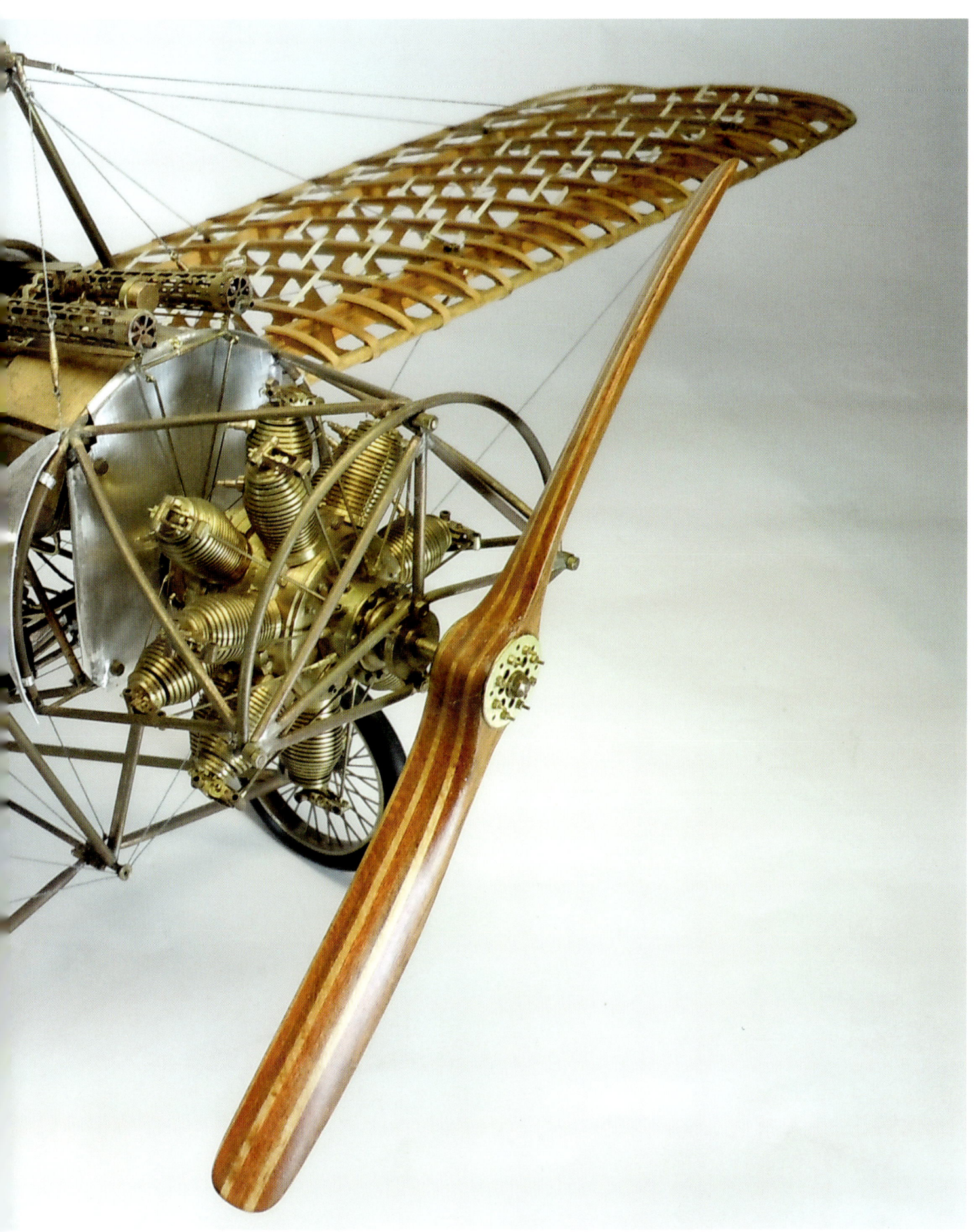

Chapter 12
Sources and Suppliers

Following are suppliers and sources for materials, tools, and equipment for modeling I have found to be very useful. Here are web addresses to their online catalogs. I encourage visiting them to learn what they have to offer and, in some cases; learn from their online tutorials about their products.

Brass Stock: K&S Engineering in store displays for brass stock.
    Online www.specialshapes.com
Micro Hardware: www.morris01550.com/contact.htm or www.scalehardware.com
Supplies and tools: www.micromark.com or www.modelexpo-online.com
Tabletop Machines: www.sherline.com
Chemicals: www.jaxchemicals.com
Electroplating: www.caswellplating.com
Reference books: *The Complete Metalsmith* by Tim McCreight and *Tabletop Machining* by Joe Martin.

Places to Visit

With the advent of the internet, access to the world of modeling is almost unlimited, with endless topics to view and sites to visit. The following are just a few of my favorites.
www.scalemotorcars.com
www.fineartmodels.com
www.modelmotorcars.com
www.craftsmanshipmuseum.com/index.html
www.wwi-models.org/index.html
www.largescaleplanes.com
www.finescale.com
www.scaleautomag.com
www.detailedmodelcars.com
www.scaleautoworks.com
www.wworkshop.net